Island

Also by J. Edward Chamberlin

The Harrowing of Eden: White Attitudes Toward Native Americans

Ripe Was the Drowsy Hour: The Age of Oscar Wilde

Come Back to Me My Language: Poetry and the West Indies

If This Is Your Land, Where Are Your Stories?
Finding Common Ground

Horse: How the Horse Has Shaped Civilizations

Island

How Islands Transform the World

J. Edward Chamberlin

First published in the United States of America in 2013 by Bluebridge, an imprint of United Tribes Media Inc.

This edition published 2013 by
Elliott and Thompson Limited
27 John Street, London WC1N 2BX
www.eandtbooks.com

ISBN: 978-1-90965-338-2

9 8 7 6 5 4 3 2 1

A catalogue record for this book is available from the British Library.

Cover design by Dan Mogford
Text design by Cynthia Dunne

Printed by TJ International Ltd.

Contents

in memory of Jack Cowdry (1921–2008)

to Rob Finley, dory compass

and for Lorna, sea anchor and heaven-haven

The natural history of these islands is eminently curious [. . .] Seeing every height crowned with its crater, and the boundaries of most of the lava-streams still distinct, we are led to believe that within a period, geologically recent, the unbroken ocean was here spread out. Hence, both in space and time, we seem to be brought somewhat near to that great fact—that mystery of mysteries—the first appearance of new beings on this earth.

CHARLES DARWIN

I should like to rise and go
Where the golden apples grow;
Where below another sky
Parrot islands anchored lie,
And, watched by cockatoos and goats,
Lonely Crusoes building boats.

ROBERT LOUIS STEVENSON

Introduction

ISLANDS ARE EVERYWHERE. There are islands in the middle of a lake, some sacred—such as Manitoulin in Lake Huron, the largest freshwater island in the world, or Isla del Sol, one of the forty or so islands in Lake Titicaca—and some sentimental, such as William Butler Yeats's Lake Isle of Innisfree or Jean-Jacques Rousseau's St. Peter's Island in Switzerland's Lake Biel; others are dear to the hearts of those who live in or visit the lake regions of the world. There are islands in rivers and streams, some supporting great cities, like New York and Montreal, others shaping cultures, like Île de la Cité in Paris, and still others whose influence seems more modest, like the "smallest, barest island" in New England's Merrimack River, which Henry David Thoreau described as having an "undefined and mysterious charm." There are islands in between, such as the Canaries and the Azores, the Hebrides and the Faroes—those so-called stepping-stone islands in the Atlantic that offered relatively safe haven to early seafarers—or the chain of outcrops called Rama's Bridge (or Adam's Bridge) that links Sri Lanka to the mainland. There are islands in the deltas of the great rivers of Asia and Africa, like the Irrawaddy and the Zambezi, and islands where land and water are confused, like the muskeg islands at the edge of the boreal forest in Canada or the Sundarbans, the mangrove swamps in the Ganges Delta (which, according to one nineteenth-century observer, "looked as though this bit of world had been left unfinished when land and sea were originally parted").

While many islands are out on the open ocean, all alone and far away from any other land—such as Tristan da Cunha in the South Atlantic and Easter Island in the Pacific—others are snuggled along the shore, like Haida Gwaii on Canada's west coast, Australia's Great Barrier Reef, and the islands that shape Singapore and Hong Kong, Mumbai and Venice. Countless rock outcrops and coral atolls are uninhabited by humans, while large islands like Java and Japan have a population of over a hundred million each.

Altogether, about one billion people live on islands. They are often fiercely (if sometimes foolishly) independent. Nearly one quarter of the members of the United Nations are island nations, some of them as small as Nauru (once called the Pleasant Island) or Tuvalu (formerly known as the Ellice Islands) in the Pacific, each under ten square miles in total and with populations around ten thousand. Eleven of the world's fifteen smallest countries are islands, including the Seychelles and the Maldives in the Indian Ocean, Malta in the Mediterranean, and several island nations in the Caribbean: Saint Kitts and Nevis; Grenada; Saint Vincent and the Grenadines; Antigua and Barbuda; and Barbados. (Their only mainland rivals for size are Vatican City, Monaco, San Marino, and Liechtenstein.)

There are islands that limit us, and islands that liberate us; islands where love flourishes, and islands where hatred takes root; islands that hold us together, and islands that keep us apart. Some islands, special for spiritual reasons, are to be visited only by the elect; others are strictly reserved for prisoners. Some, with material resources, have been occupied by a few families for centuries, while other islands, with no resources at all, are now home to thousands of residents.

People have gone to war over islands, as they did with the Falklands and the island of Run (now part of Indonesia), which was the only source of the precious spice nutmeg during the seventeenth century. And islands have been instrumental in making peace: the British ceded the very same Run to the Dutch in 1667 in exchange for Manhattan, and France traded its part of Canada (and more) to Britain in 1763 in order to secure Guadeloupe and Martinique—because of the islands' sugar cane.

Many islands stay put, like sentinels of the sea and guardians of the straight and narrow, and others move about with the wind or the current or the quirks of fate, like the Flemish sandbanks and Sable Island (off the coast of Nova Scotia) and the floating islands of roots and vegetation in the Florida Everglades and in the Tigris–Euphrates Delta. Some islands disappear and reappear—the Halligen islands in the North Sea during stormy season, Falcon Island in the Pacific once in a volcanic blue moon—while others can be reached on foot, but only at low tide. And there have been man-made islands for thousands of years, from the crannogs in ancient Ireland and Scotland to the prehistoric villages built on stilts in Alpine lakes.

There are islands we escape to—and islands we escape from. Some of them are real, and some are imagined. When mapmaking became a cultural tradition—especially in Europe and Asia—as well as a travel guide, plenty of imaginary islands appeared on these "real" maps. Commerce played a big role in this. No one ever landed on the mythical island of Buss in the North Atlantic, for example, but it was still charted on maps and even chartered to the Hudson's

Bay Company to harvest furs. Such islands are both there and not there—like stories. "It was, and it was not" is the phrase used by storytellers on the island of Majorca when they begin. Maybe stories themselves began with islands, for islands have fascinated people as long as they have been singing songs and telling tales and traveling, and have found counterparts in the islands that are our homes and gardens and towns and farms, as well as our personalities. For millennia, seafarers and settlers and storytellers have sought out islands for reasons that go deep into the human psyche and haunt its imagination, even—or sometimes especially—when ignorant of geography. It may have something to do with the way an island rises up from the sea and then sooner or later disappears again, perhaps invoking a primordial consciousness of the beginning and the end of life. Or it may be connected to the journey between the mainland and an island, and between one island and another, requiring the crossing of water. This has haunted humans since time immemorial; the word "metaphor," the signature of stories and songs, means "to carry across."

Islands have provided a special invitation to fertile imaginations, just as they did (in evolutionary theory) to unique mutations. There have been countless islands where marvelous—or malicious—things supposedly happened, and memorable islands that came into being as fiction took up history, with stories of true island adventures (survivor stories long before reality TV) sponsoring novels like *Robinson Crusoe*. Psychiatry began with an awareness of "islanded" psyches, and anthropology made islands an academic fetish (and a travel excuse), with island accounts beginning in the late nineteenth century by Arthur Haddon on the Torres Strait

Islands and Alfred Radcliffe-Brown on the Andamans, and then by Bronislaw Malinowski on the Trobriand Islands and Margaret Mead on Samoa.

Islands represent both paradise and purgatory, just as they invoke madness and invite magic. They have been places where curious things occur—or where nothing at all takes place. But even then, the howling noises of the sea or the deep silence of a lake will often conjure up a sense of strangeness around islands, and generate stories about the unusual things that go on there. Many poets, from the Scandinavian *skald* to the Swahili *sha'ir* and from Homer (in the *Odyssey*) to Shakespeare (in *The Tempest*), have located some of their most intriguing stories on islands. Later, Jonathan Swift took readers to islands of wonderment on Gulliver's travels, Alexandre Dumas to the treasure of Monte Cristo, and H. G. Wells to the menacing *Island of Dr. Moreau*. And they are certainly places where fabled creatures live: there are real islands with dragons, like Komodo in Indonesia, and imaginary islands with dragons, like those in the *Chronicles of Narnia*.

There are islands of solitude, and islands with a social life—though not always an easy one. So is the ultimate appeal of islands "home"—or "away"? Ideals of civilized life, domestic and settled, are routinely represented in island images, but so are concepts of the wild and the barbaric; and both of them may sustain the sense of community that islands often celebrate.

The history of islands is also the history of our planet, from its beginning as an island in space to its current position as part of the archipelago that is our solar system, and from the moment land first appeared above the waters that covered the earth to the contemporary appearance and disappearance of islands in the cycles of climate change and seismic upheaval that make up and break up our world.

All of which raises—or complicates—the question: What is an island? Is it simply land surrounded by water, which the etymology of the word in various languages suggests? Do tidal islands, and isolated peninsulas, qualify? How about man-made islands, like oil rigs or waterfront real estate developments—or castles surrounded by moats? Is size a factor, with small being beautiful? But then, what is it that a reef or a rock outcrop have in common with Greenland or New Guinea? And what about continents like Australia and Antarctica? Do geology and geography set the standard for island identity, or politics and economics? Are islands defined by their natural history—or by their human history?

One thing is certain: barren or beautiful, large or small, real or imagined, islands are a central part of the world we live in. They represent much of what we dread, and much of what we desire. And since so many of our thoughts and feelings have an island counterpart, they may well define what it is to be human.

Chapter One

First Islanders

Settlers and Storytellers

"Jamaica, the most considerable as well as by far the most valuable of the British West India islands, is situated in the Atlantic Ocean, among what are called by geographers the Greater Antilles [. . .] Jamaica is nearly of an oval form; 140 English miles in length, and in its broadest part about 50. It is the third in size of the islands of the Archipelago. It is bounded on the east by the island of St. Domingo [Hispaniola], from which it is separated by the channel called by English seamen the Windward passage; by Cuba on the north; by the Bay of Honduras on the west; and by Cartagena in New Spain [now Colombia] on the south [. . .] The island is crossed longitudinally by an elevated ridge, called the Blue Mountains. What is called the Blue Mountain Peak rises 7,431 feet above the level of the sea. The precipices are interspersed with beautiful savannahs, and are clothed with vast forests of mahogany, lignum vitae, iron wood, logwood, braziletto, etc. On the north of the island, at a small distance from the sea, the land rises in small round topped hills, which are covered with spontaneous groves of pimento. Under the shade of these is a beautiful rich turf. This side of the island is also well watered, every valley having its rivulet, many of which tumble from overhanging cliffs into the sea. The background in this prospect, consisting of a vast amphitheatre of forests, melting gradually into the distant Blue Mountains, is very striking. On the south coast the face of the country is different; it is more sublime, but not so pleasing. The mountains here approach the sea in immense ridges; but there are even here cultivated spots on the sides of the hills, and in many parts vast savannahs—covered with sugar canes, stretching from the sea to the foot of the mountain—relieve and soften the savage grandeur of the prospect."

—*Edinburgh Encyclopaedia* (1830)

"[December 4, 1844] My first sight of Jamaica was one that I never can forget. There was a conical mass, darkly blue, above the dense bed of clouds that hung around its sides, and enveloped all beneath [the] towering elevation [of Blue Mountain Peak] [. . .] Night soon fell. Many lights were seen in the scattered cottages, and here and there a fire blazed up from the beach, or a torch in the hand of some fisherman was carried from place to place. My mind was full of Columbus, and of his feelings on that eventful night, when the coast of Guanahani [San Salvador, also known as Watlings Island, in the Bahamas] lay spread out before him with its moving lights, and proud anticipations. So did I contemplate the tropical island before me, its romance heightened by the indefiniteness and obscurity in which it lay [. . .] The well-known comparison by which Columbus is said to have given Queen Isabella an idea of Jamaica—a sheet of paper crumpled up tightly in the hand, and then partially stretched out—occurred to me, and I could not but admire its striking appropriateness."

—Philip Henry Gosse, *A Naturalist's Sojourn in Jamaica* (1851)

Like many islanders the world over, Jamaicans often refer to their island as the Rock; and in the beginning Jamaica was just that, a rock rising above the surface of the sea. Long after its geological history had begun, its natural history followed, with biological life forming in the soupy sea around while on the land plants developed over millions of years until eventually the first tree rose up into the air.

Jamaica would one day be covered by trees; but before that it was shaped by volcanic rocks, which were eroded by wind and water and by limestone produced with the slow disintegration of the shells of marine life, including coral. This limestone then dissolved into pockets and purses called karst formations, the sinkholes and springs and caves and underground reservoirs and cone-shaped hills and honeycomb sides of Jamaica's Cockpit Country, one of the most remarkable examples of such a limestone landscape in the world.

Then, in a familiar island compact, Jamaica's geography conspired with its geology to give the island a rich diversity of flora and fauna, fascinating travelers and residents alike from ancient times. Such variety remains one of the most reliable markers of island identity. Temperate or tropical, on almost every ocean island around the world there are thousands of plant and animal species, some tiny and hidden away and others portly and prominent, some resident and others just passing through; and over time they create their own island conditions, building a world in which they uniquely belong and bringing nurture and nature together in what Jamaicans might call a rocksteady rhythm. Typical of mountainous islands with windward and leeward shores, Jamaica has two climates; in the northeast, exposed to the wind, it is wet and warm, while in the lee of the mountains to the southwest the weather is drier and cooler. This makes for an even more remarkable range of species, including an exceptional variety of ferns and fruits and berries—and of birds to feed on them.

Birds and islands have a long association, for until very, very recently in the history of the world, birds were the only creatures (other than insects and bats) that could *fly* over the water to reach islands. The Chinese word for an island combines the ancient ideogram of a bird with that of a mountain—and measured from the seabed, every ocean island is a mountain. In this imagining, an island is a place for a bird to land; and birds hide out or hover about all over Jamaica—it has the highest number of native avian species anywhere in the Caribbean. There are finches and flycatchers, herons and egrets, black-billed and yellow-billed parrots, mockingbirds and kingbirds and the magnificent frigate bird, with its scarlet throat that balloons in breeding season. There are whistling ducks and blue and yellow and black and white warblers, elaenias (called Sarah Birds) and euphonias (called Cho-Cho Quits), an owl called Patoo and an oriole called Banana Katie. The island is also home to a little green and red tody called Rasta Bird, a cuckoo called Old Woman Bird and another called Old Man Bird, and to doves and lots of crows, including the rightly named jabbering crow. And the John Crow, which isn't really a crow at all but a turkey vulture (graceful or disgraceful, depending on one's point of view) that has found a place in Jamaican folklore. There are hummingbirds, especially the so-called Doctor Bird, found only on the island and celebrated in Ian Fleming's James Bond story *For Your Eyes Only* (1960), where the opening words speak of "the most beautiful bird in Jamaica, and some say the most beautiful bird in the world, the streamer-tail or 'doctor' hummingbird." It is now the national bird of Jamaica. And there are over a hundred species of butterflies, each of exquisite color and design, including the giant swallowtail—largest in the Americas and native to the island.

Green and brown and croaking lizards live in Jamaica, along with a few snakes and one species of crocodile (which appears on the national coat of arms). Turtles have lived on and around Jamaica for millions of years, from the time when the sea was relatively shallow; and now there are green turtles and hawksbills and loggerheads and leatherbacks, bringing the sea and the land together in their amphibious lives. In fact, certain reptiles that can manage significant water crossings are usually among the first animals on any "new" island. Perhaps as importantly, reptiles can go for long periods without eating anything at all, absorbing heat passively from the sun and the ambient air rather than having to generate body heat internally like mammals and birds—which requires a lot of food.

Among the marine mammals, the manatees, or "sea cows," are relatively scarce around Jamaica these days; but they have made their home close to shore forever, and would have been easy pickings for the earliest human settlers. There are still plenty of dolphins and some whales, including sperm whales and humpbacks, sharing the waters with stingrays and sharks, barracudas and eels, marlin and tuna. Around the reefs there are fish with intriguing names—grunts and groupers, snappers and doctorfish, squirrelfish and goatfish and triggerfish and angelfish—and in some parts of the Caribbean Sea the unusual flying fish, breaking the surface at speeds up to forty miles per hour and gliding on pectoral and pelvic fins for distances from a hundred feet to a quarter mile. Some say it flies over the waves to escape predators; others believe it does it just for fun.

The trees that originally grew on Jamaica, from the dry lowlands to the rainy valleys and up onto the steep hillsides, were drastically reduced—in both number and variety—with the arrival of humans, who cut timber for building and cleared land for agriculture. But this opened up spaces for new plant varieties, some introduced by travelers and some brought there by chance, on the winds and waves or by bird or boat. There are tamarinds whose seeds traveled from Asia, breadfruits brought from Tahiti by William Bligh (the notorious commander of the *Bounty*), and sapodilla (better known in Jamaica as naseberry) with its delicious fruit and sap called chicle, tapped every six years to make chewing gum. Jamaica has flowering trees such as the native poui, which has a spectacular yellow-gold blossom that lasts only a short while, falling in an apron around the base of the tree; and the blue mahoe, which displays trumpet-shaped, hibiscus-like flowers at different times of the year, its wood prized for cabinet-making and the crafting of musical instruments, and its bark used in Cuba to wrap cigars.

Other trees have various uses. The lignum vitae's sapwood was once a remedy for syphilis, and its gum is still used in the treatment of arthritis. There is a tree called Duppy Machete, with floral petals deemed suitable for dealing with duppies, West Indian spirits of the dead. The beautiful frangipani has a sweet scent but a poisonous sap, while the fruit of the ackee is poisonous at the wrong time of harvesting but delicious at the right stage. It is now the foundation of Jamaica's national dish, ackee and salt (cod) fish. There are cashew trees, with a pear-shaped fruit that ends in a kernel that is the nut, and soursop and coconut and pawpaw and mango trees, along with avocado and banana and cocoa and nutmeg, none of them native but all now widely dispersed on the islands of the Caribbean. Nutmeg was an icon of the spice trade, which once centered around a group of islands in Indonesia but was eventually transplanted to Grenada and other West Indian islands. The berries of the indigenous pimenta tree seemed to early European travelers to combine the taste of nutmeg, cinnamon, and clove, and for a long time Jamaica supplied most of the world with allspice, as it was called.

Mangroves grow on many of the Caribbean island shores, as they do in many parts of the world, extending the reach of islands by walking out to sea with their prop roots and capturing sediment and plants that will eventually shape the swamp into an expanded shoreline. There are casuarinas, which have come from afar but made the Caribbean their home, and sea grapes, with their leathery leaves and sour-grape fruit. They may have been the first plant seen by Columbus when he reached what he thought were the Spice Islands of Indonesia—and the smell of these trees, along with the sight of their leaves blown from the shore and carried by the sea, would have been noticed by sailors long before any island came into view.

Anthurium, bougainvillea, Easter lily, wild scallion, and heliconia—called wild banana—are among the flowering plants native to Jamaica, along with a mimosa called Shame-Me-Lady, so sensitive that a light touch or a slight breeze cause the leaf stalks to collapse and the leaflets to close. One indigenous plant called Ram Goat Dash Along makes a healing bush tea, while cerasee makes a bitter tea that is also used as a body wash. Pomegranate and frangipani shrubs are common, along with Duppy Cho-Cho, which may harbor bad spirits. Marigolds and fuchsia grow in the mountains. Jamaica is also home to the greatest variety of orchid species in the Caribbean. They originated in Africa, like the enslaved men, women, and children who were later transported to the island, but many of the orchids were brought to the Americas as seed dust on the Sahara winds.

All humans on the islands of the world are settlers, though when they first traveled there, and why and how and from where, is often uncertain, or else explained in the myths that make up the history of first islanders (and which usually include stories about the first plants and animals as well). We can be sure that some humans came by choice, some reached by chance, and coercion played a part for others. Often islands provided sanctuary for those fleeing hunger or war, or seeking solitude—saintly or otherwise. Islands saw the arrival both of seasonal workers and of enslaved laborers, and of settlers looking for a different life, establishing new societies, and exploiting natural resources that they did not have back home. And some of them will have had a dream.

People seem to have first settled the islands of the Caribbean around six thousand years ago. All of them came by boat—some paddling, some perhaps sailing, others just drifting from the mainland—though over time a myth was told about flying to the islands, transported by birds or spirits. Once there, they gathered wild plants and ocean kelp and hunted food from the seashell-crunchy, seaweed-squishy shore; and according to their stories they were only the latest in a series of travelers in the Caribbean Sea stretching back into the mists of time. Alternating periods of wandering and settling down had defined the lives of these Amerindian people—indeed of all people—since the beginning, with each coming and going being different and yet the same, signaling both a passage and a pattern. Some set out from the mouth of the Orinoco River in Venezuela, and their first islands were Trinidad and Tobago. Others came from Central America along the Yucatán Peninsula and from Florida, and they settled on Puerto Rico, Hispaniola, Cuba, and Jamaica (often referred to as the Greater Antilles). Archaeologists have divided them into different indigenous groups, but such labels are misleading, for they all thought of themselves simply as The People. "We the people" is the quintessential island affirmation, for an island is not a metaphor for home. Home is a metaphor for an island.

We can only guess whether it was because of a crisis or out of curiosity that these ancient peoples began to set out to sea from the American mainland. They may have had dreams of a place where they would find material and spiritual well-being; or maybe they had actually heard about such a place from their singers and storytellers.

Some of them may have been looking for a new start; others might have just gone for a boat ride and lost their way. Perhaps they were on their way to meet their ancestors. Whatever their motivation, they *did* set out.

When these first Amerindians arrived on Trinidad and Tobago, they couldn't see anywhere else to go and so they stayed, imagining these isles for awhile as the new center of their world. Then they heard from their dreamers and derring-doers about other islands, far beyond the horizon—about one hundred miles across the sea, as it happened. So when they had had enough of life on those first islands, or enough of their fellow islanders, or were restless for adventure, they set off again, paddling and drifting until they reached the island of Grenada. From there they could see another island, and another, and another, part of an arc stretching some five hundred miles—the eastern Caribbean (also called the Lesser Antilles). Were these the blessed isles their shamans had spoken about, over the horizon, at the end of the rainbow? Were they the home of spirits of malice and mischief or of gods of grace and goodness? Nobody knew. And everybody wondered.

These ancient Amerindian peoples were used to the ways in which rivers and mountains offered plants and animals to them, and as they traveled north along the arc of islands, they continued to harvest some of the shore food they were already familiar with. But fishing in the open sea offered another livelihood, where hunting and gathering required new knowledge and skills and a surrender to different natural and supernatural forces; and these soon became part of their consciousness and their culture. Over time, they made these islands of the archipelago their new home. They found oysters, mussels, conch, and crab along the shoreline and in the mangrove swamps, and larger species—including lobsters two feet long and weighing over thirty pounds—on the sand, among the sea grass, and on the rocky beaches. They took to the sea for fish, and they ventured inland, finding some animals and plants they knew about and others they had never seen before. They harvested birds and reptiles and the small mammals that had swum to the islands or stolen a ride on driftwood. Slowly they brought their hunting and harvesting heritages into harmony, with island birds and sea turtles now animating their myths, and island storylines telling about their new relationship with the land and the sea around it.

Still, the history of settlement in the Caribbean was far from over. It seldom is with islands, where comings and goings are facts of life. From the South American mainland, new settlers with new ways of living began arriving in the Caribbean around 500 BCE, moving throughout the islands. They cultivated crops and resided in communal dwellings and village centers rather than seasonal hunting and fishing camps. They built houses to last for generations, farmed the land, harvested the sea, and created sophisticated ceremonies. Because many of the islands are mountainous, some of these new Amerindian settlers established political strongholds in the highland interiors where the resources were plentiful and the competition scarce. They expanded the existing traditions of weaving and basket-making and ceramics and developed forms of dance and music and cooking that caught the attention of the Europeans and Africans who came much later; and they became known as the Arawak—from *aru*, their word for cassava.

Over time, the culture of the Arawak developed in distinctive ways on different islands of the Caribbean. This, too, is the story of life on islands all over the world. (When Charles Darwin visited the Galápagos archipelago, he was intrigued by the differences in flora and fauna on islands only a few miles apart.) And so on the island of Jamaica unique human cultures and languages emerged, with arts and crafts and games that surprised other islanders, even those relatively close by. A system of chiefdoms provided political stability, inheriting power along the matriarchal line (which seemed unnatural to the European seafarers and nostalgic to the Africans when they came as enslaved laborers), and a class system that allocated responsibilities and obligations according to rank. Shamans brought supernatural resources to everyday life, including the medical and the military, inhaling an hallucinogenic powder for healing and holy enterprise called *cohoba*, ground from the seeds of a local tree. Perhaps their most intriguing artifacts were ceremonial seats (called *duhos*) and triangular carved stones—a sculptural expression of their spiritual concept of *zemis*, which were symbolic of ancestral and cosmic power.

The people throughout the Greater Antilles eventually became known as Taino, and their stories provided new island cosmologies and new understandings, both scientific and religious, of natural forces like the hurricane, a Taino word. They refined technologies such as the canoe and the barbecue and smoking tobacco, words that also come from the Taino language. They called themselves *lukku cairi*, which means island people. By the time the Spanish arrived in the late fifteenth century, several thousand Taino lived on the island of Jamaica, its harbors home to seafarers, its valleys busy with agricultural and ceremonial activities, and its mountains providing hiding places and safe havens. The distinctive character of their culture was immediately obvious to the Spaniards and persuaded the perceptive among them that this was a complex and sophisticated civilization.

Taino culture continued to inform life after the arrival of the Spanish and other European settlers and then of the enslaved Africans, though as a separate people the Taino lasted only about a century before succumbing to disease, the social and economic depredations of the sugar cane agribusiness, and both friendly and unfriendly intercourse with the Europeans (with whom they shared a fondness for accumulating material as well as spiritual resources, and a facility for separating labor and capital). The arrival of enslaved Africans created island societies different not only from the African homelands which they had been forced to leave but also from the societies that were developing on the mainlands of North and South America; and these differences, though often masked by the shared experience of slavery, are still apparent, signaling something important about the way island life distinguishes itself from its continental counterpart.

The story of European and then African settlement in the Caribbean is in most ways a grim one, and has been often told. It is the story of people in a hurry to make money—which is another typical island story, it turns out, with versions around the world that chronicle the human lust for power and privilege and for products as varied as sealskins and spices and sex. In the case of the Caribbean, it was sugar; and the slavery that made sugar production possible, and for a while highly profitable, deeply warped its island life.

Christopher Columbus did not have all this in mind when he first landed on the islands of the Caribbean. He was completely exhausted, probably a bit scared, and more than a little surprised. First visiting Jamaica on his second voyage in 1494, and still convinced that he had found the Spice Islands of the East, Columbus described the island as "the fairest that eyes had beheld." It was also "mountainous and heavily populated," he noted, rather nervously, since he wanted to deal with the indigenous people, whom he had described in his earlier "letter announcing the discovery of the new world" as "of a very astute intelligence, and they are men who navigate all these seas." The Taino of Jamaica were gentle and generous in their own territory, but jealous of islanders elsewhere. They disliked and distrusted the Amerindian Caribs in the Lesser Antilles and persuaded Columbus that the Caribs were cannibals. They weren't, at least not in the way the Taino suggested; but that hardly mattered in the jostle of island jealousies. Just so, contemporary Jamaicans hold a special disregard for the "small islanders" of the eastern Caribbean, describing them in all sorts of ways that are, let us say, unflattering. And of course those islanders return the favor.

This is an island habit, it seems; or maybe it is a human one, though perhaps only mountain societies, bound by dialect and cultural differences, display the same capacity as islanders for prejudice against communities only a short distance away but separated by seemingly impassable barriers. Such insularity shows itself in particular ways on certain islands, reinforced by a defensively conservative attachment to traditional practices and ceremonies as well as by ingenious social and economic arrangements, which may explain why the English word "insular" took on negative connotations over time.

But also on islands, something positive is often involved. The

Amerindian peoples, arriving separately but finding themselves sharing islands in the Caribbean, had to find new ways of interacting—and new social, economic, political, and cultural dynamics that combined elements of both their hunter-gatherer and their agricultural communities. What emerged after centuries were unique multicultural and multilingual societies, rare at any time and exceptional in the long history of human interaction. On the mainland, the story of agricultural societies impatiently destroying hunter-gatherer communities is the norm; but on some of these islands, despite—or maybe because of—the relatively close quarters they were confined to, the various settlers reconciled these livelihoods, drawing on spiritual as well as material technologies from each and combining hunting with harvesting on land and sea and shore.

For other than setting back out to sea, or flying away on the wings of a dove, there was nowhere else for these people to go. So they could either destroy each other, or find ways of getting along. Anything is possible given time, and on islands time moves at whatever pace people want. Which means that when people are in a hurry, terrible things tend to happen. When they are not, islands can be places of peace and tranquility, or at least of reconciliation; for like everywhere on Earth, in geology no less than in the natural world, island life is a work in progress, and from time to time islanders have nourished remarkable ways of settling their differences. It may have had something to do with the water all around, for as the pioneer ecologist Rachel Carson said in her book *The Sea Around Us* (1951), there is no "more delicately balanced relationship than that of island life to its environment." Which includes other human beings. Iceland and the Isle of Man are home to the world's oldest parliaments, after all, and the parliament in England also has a good long history, while the story of ancient Polynesian political culture is, mostly, one of remarkable checks and balances. ("A difference of opinion surrounded by water" is how one islander describes where he lives on Salt Spring Island on the west coast of Canada.) It doesn't always work out, of course; but willy-nilly, the sea concentrates an islander's mind.

The "i" in the English word "island" comes from the Anglo-Saxon *eig* and the Old Norse *ey*, which mean water. And water, which covers nearly three quarters of our planet's surface and surrounds all the continents, is a place where we don't belong. The various myths about floods, common to peoples all over the world, are a reminder not only of the occasionally catastrophic consequences of climate change (or, in some accounts, the grumpiness of the gods), but of the perpetual condition of humans in relation to water.

Water is a paradox. It is a source of both life and death for us; and for some ocean islanders, especially, water is something that unites rather than separates, its bountiful blessings more important than its occasional brutality, a place of peace that passeth understanding. On the island of Tikopia in the Solomon Islands, death at sea is referred to as a "sweet burial." When the islander William Wordsworth, in a time of trouble at the beginning of the 1800s, called on John Milton for comfort, he said the great seventeenth-century poet "had a voice whose sound was like the sea." Life began in the sea, as far as we know; and when we seek evidence of life on other planets we look for water. But no matter how friendly our feelings toward it, we are not at home in the water. We don't swim very well, or for very long, certainly not compared to our cousins the marine mammals, much less the fish and reptiles that live there. And while small amounts of salt water are cleansing, seawater makes us sick. "You can't drink the sea," sings the Newfoundlander Ron Hynes, echoing Samuel Taylor Coleridge's ancient Mariner who cried "water, water, everywhere, nor any drop to drink" shortly after he shot an albatross, perhaps not even knowing that albatrosses drink seawater quite happily.

The sea is undomestic, wild. "No man can ever in truth declare that he saw the sea look young, as the earth looks young in spring" observed Joseph Conrad, and the Newfoundland poet E. J. Pratt wrote: "There is no silence upon the earth or under the earth like the silence under the sea; / No cries announcing birth, / No sounds declaring death." "Nothing can be put down in the sea. You can't plant on it, you can't live on it, you can't walk on it," reflects the Nobel laureate Derek Walcott from his island of St. Lucia. The sea "does not have anything on it that is a memento of man," he adds; and memory is the mark of humankind.

The ocean is the only domain, other than outer space, where humans are so completely alien and where wonder holds us so close. This is the heart of the matter, for it is this wonder that has inspired voyagers for millennia to row and sail the seas in search of an island, sometimes any island; and it is this wonder that still inspires us to look for a planetary island like ours in the furthest reaches of space. Ultimately, we are all islanders on planet Earth, surrounded by the air and the water which we need in order to survive but which by themselves will not sustain us.

"No matter which direction I walked I would arrive at the border of another wilderness, the savage sea," recalls the writer Thurston

Clarke of his visit to Más a Tierra, one of the three Juan Fernández Islands off the coast of Chile. Más a Tierra, which means "close to the mainland" (four hundred miles away!), is the island where Alexander Selkirk was marooned in 1704; and Selkirk's four-year sojourn there was the inspiration for Daniel Defoe's classic island novel, *Robinson Crusoe* (1719). "An island wilderness is different and more perfect than a continental one," Clarke continues. And indeed there is something different—no, something *in*different, inhuman—about the sea. Only death is as indifferent, which is the sentiment behind the poet John Donne's famous line: "No man is an island, entire of itself; every man is a piece of the continent, a part of the main." He wrote this in a prose meditation on death in 1624, and its truth also lies in the contradiction that the separateness of islands as well as of humans is underwritten by their respective connectedness. Not unlike the ocean floor, which is shared by sea islands and connects them underwater, our human connection is language, language that reminds us of both how united and how divided we are. In the end, the connection lies in death, which is to say in our shared human mortality, just as all islands will disappear eventually.

It is sometimes said that language is what defines us as humans. But it is really belief, and ceremonies of belief, of which language may be the most remarkable. It is not the only such ceremony, however. Just as language—its words and images—requires us to believe in its artifice, its man-made (or divinely inspired) ability to take us across the gulf that separates us as individuals, so leaving the shore and sailing to an island involves an act of faith in the technologies of craft and navigation.

"The water is wide, I cannot cross over, neither have I wings to fly. Give me a boat," begins a famous lament from the British isles. Separateness is a condition that fills humans with both dread and delight. Much ancient and modern philosophy, politics, and now economics are about the insularity of our individual consciousness and the ingenious ways we have developed of making connections, forming relationships, and establishing commerce between you and me or them and us, while also maintaining distance and difference. "Thank God we're surrounded by water" is the chorus line of a song by Dominic Behan (Ireland) and Tom Cahill (Newfoundland), celebrating the advantages of being separated from others by the sea. "Thank God we speak Irish" (which is to say, Gaelic rather than English) is its cultural counterpart, for we are all islanded by our individual languages. "Shades of the prison house begin to close upon the growing boy," said Wordsworth about first learning a language. Walter Pater, writing in his book on the Renaissance (1873) about the inner and outer worlds of consciousness that words and images bring together, described how we can be bound by the very intelligence and imagination that give us freedom, with "each mind keeping as a solitary prisoner its own dream of a world."

The contradiction is always there. Both islands and languages lay claim to their inhabitants, limiting as well as liberating them, holding them hostage even as they set them free—though from what, people don't always agree. Indeed, it may be a consciousness of our "islanded" existence—and our capacity (and, more often than not, desire) for crossing the wide water between—that makes us truly human, a consciousness and a capacity that incorporates our uniquely human understanding of life and death, the ultimate separation. Crossing the water is an ancient image for the passing over that is death, as Donne knew well, and it is an image shared among many religious traditions. The Christian community, for its part, is often likened to a ship, its church nave and other architectural elements modeled after the ark that saved Noah's family, with the priest as navigator and the cross conflating mast and anchor. "No man is an island" is a prayer as well as a proposition.

In the British Museum is an ancient Taino sculpture from Jamaica that portrays a bird on the back of a turtle. It represents a creation story that has wide currency among indigenous peoples in the Americas, telling of a special tree that grew on an island high above the world, and an ancient chief who lived there with a woman who was his beloved. She had a dream that the tree had been uprooted, and when she awoke she told her man about it. They went to look, and there it was still standing where it had always been. Just a dream. But dreams must be taken seriously, and the chief decided he'd better do something to make it come true. So he pulled the tree out of the ground.

In art as well as life, an action like this is usually followed by a reaction. Something is given, something must be taken away. Something is lifted up, something must fall down. And sure enough, the tree that the old man had uprooted left a great big hole; and when the woman came to look at it, she fell through.

Down below, water covered everything. The only living things were the fish and the seafaring animals and the birds. They all looked up, and saw a woman falling from the sky—like a meteor, which she eventually became in the stories of science. To save her, two seabirds—swans, some say, or maybe cormorants—caught and balanced her on their wings. They flew about for a long time, but eventually they needed somewhere to rest. Except there wasn't anywhere. Just the sea below and the sky above.

Another of the waterbirds said she had heard that there was earth far below the surface of the sea. That would be ideal, they all agreed; but how to get it, if indeed there was earth down there. Everyone offered to help. First a beaver went down, but he didn't find any. Then a loon tried, going down and down, but he, too, came up empty. Finally, a muskrat gave it a go, diving deeper and deeper until, just when she could go no further, she reached bottom, grabbed a pawful of earth, and swam back to the surface, gasping for air.

But where to put the earth? A turtle, swimming by at that very moment, said, "put it on my back." Which the muskrat did. And birds had a place to land. Trees had a place to grow. Humans had a home called Turtle Island. And the whole world as we know it came into being.

In other accounts of the origin of the earth, a bird drops dirt onto the back of a whale, or onto a mythical water creature. There is a Taino legend about a sacred calabash that contained all the fruits of the sea. It was stolen and then dropped, the water in the calabash flooding the earth, the only parts spared being the mountains that form the islands of the Caribbean Sea. A story from the Pacific islands explains how a boat carrying people—an ark, of sorts—is turned into an island, and then the people call on Katinanik, the mangrove, to protect it from the waves, and on Katenenior, the barrier reef, to surround it. The Maori of New Zealand tell of a bird lifting the land out of the water. In a story from Hawaii, a bird lays an egg that is fertilized by the sun. Still other stories describe how a bird makes a place to land out of twigs and branches, building up an island the way science describes it happening with volcanic ash and the accumulation of sediment, and the way humans have been making artificial islands in rivers and lakes for millennia. Many Polynesian legends include islands being brought to the surface on a fishhook.

On one island archipelago in the North Pacific, people tell of the time when a loon swam about for days and days, looking for land. He'd seen a cloud in the sky, so he flew up and there he found a dwelling in which an old man was lying beside two quartz stones, burning bright. The old man didn't move, and so the loon went outside and cried, making the call he still makes today. He cried all afternoon and all night and all the next day—until on the third day the old man woke up and complained that he couldn't sleep with all that noise. The loon explained that he was crying because there was no place down below for the people to live. So the old man gave the loon a small black stone that he took from a box within a box within a box, and he told him to go down and place it in the water and breathe on it for a short while. And then he gave him a large stone with shiny things running through it and told him to do the same, but to breathe on it for as long as he could; and the small stone became Haida Gwaii—the archipelago once known as the Queen Charlotte Islands—and the large stone the continent of the Americas. The old man then created a raven to fly down and land on the islands; and later the real people came out of the sea.

That islands should be part of so many creation stories is hardly surprising, for they provide an image of that first moment when the land was separated from the waters and human life was made possible. In stories of the end of the world, too, islands are usually the last place left. Like floods, earthquakes, and volcanoes, islands are often associated with powers or forces beyond our comprehension but fundamental to our understanding of the world. Some of these forces are imaginary (or personified in the gods who are believed to inhabit the world), but they can seem very real. Myths are often misunderstood as distancing us from such forces. In fact, they make them *more* real, not less. And the power of such myths comes not from the fact that fires and floods, for instance, are common across cultures, but from the way the stories about them bring together scientific and religious accounts, allowing each its own authority without discrediting the other. We believe such stories not only to make sense of the world, or to take control of it, but also to remind ourselves that some things don't make sense, and some things we can't control. The stories of religion show us how to accept these forces. The stories of science show us that we don't always have to. Creation stories that begin with birds and turtles and firestones and fishhooks are not mistaken explanations of historical incidents but true explanations of the human condition and of our very human wonder about the mystery of creation and destruction. Which is the mystery of islands.

The first bird mentioned in the Bible is a raven, related by more than ornithology to the one in the Haida Gwaii story. Noah, adrift on the waters, sent the bird out and it never came back. Seemingly no comfort there. Except that there *was*, for Noah knew something about ravens from stories that were told back then, stories in which ravens look after themselves; and he realized that this raven must have found a place to rest. An island, of course, since there was nothing else. No sign of the raven was, for Noah, a hopeful sign. But like a good scientist, he needed some evidence. So he sent out a dove, a homing pigeon. They always come back, he'd been told; and sure enough it did, carrying a leaf. That was enough. An island with a tree on it. A place for a bird to land. Noah had proof that the floodwaters were receding and islands appearing once again; and one of them, which much later became known to us as Mount Ararat, provided a resting place for the ark.

In some stories of the beginning of the world, nothing came first, nothing at all—which is why the opening of John's Gospel, in its early Greek version, has no definite article. "In beginning," it begins. But all the stories agree that sooner or later *something* happened—a word perhaps, or a deed. Many storytellers hedge their bets and hold out for both. Whatever the case, beginning is marked by a difference: a new note in the scale, a new star in the sky, a new color on the canvas. A bang or a whimper. A line separating above from below, light from darkness. A spot of space or time. And, more often than not, an island.

Most of them offer an image to remember that moment—fire or flood, bird droppings or egg hatchings, island craftings or ocean crossings, fallings or risings. Science and religion collaborate on this more than we realize, weaving together stories that have been told around the world forever. Sometimes they describe events that happened in historical time, sometimes in the mythical past, and often it's impossible to tell the difference. And the question we love to ask—whether it all began with land or water, island or ocean—is like asking which came first, the chicken or the egg. A chicken may simply be an egg's way of producing another egg. Science and religion dance around this like revelers around a beach fire—and whether about rocks or ravens, tectonic plates or turtles, that's where creation stories come into their own.

Philosophers were the popularizers of science in ancient times, and their speculations had the same authority we now accord scientific accounts. In fact, "philosopher" as the name for a scientist had currency up until late in the nineteenth century, preferred by naturalists as notable as Charles Darwin and Thomas Huxley. The ancient philosopher-scientists proposed a variety of early representations of the world, and again and again they envisioned land encircled by water. In the sixth century BCE, the Greek philosopher Anaximander, perhaps influenced by images from the Middle East, described the earth as a disc floating in water; and images of the earth as a circle, a square, or a disc, often with some sort of roof or umbrella overhead and sometimes surrounded both above and below by water, were widely circulated in Europe and Asia and North Africa. Around the same time, Pythagoras argued the case for a round planet on philosophic grounds, because a sphere was the most perfect shape and motionless (movement could be undignified). He was a mathematician, after all, and mathematicians like elegant explanations. In the fourth century BCE, Pythagoras's ideas were picked up by Plato and written into scientific scripture by Aristotle. The Roman Macrobius, writing around 400 CE, suggested a geocentric model of the cosmos, with the earth in the middle surrounded by water and air and fire as well as a set of four planetary islands figured as a quadrille. The earth as a sphere was routinely represented in popular globes that were made and marketed in Europe as early as the thirteenth century. And an Islamic map titled "The Wonders of Creation" from the sixteenth century CE showed the earth surrounded by the sea, which was in turn surrounded by a mountain, all nested in a bowl of water.

Plato also described an island continent he called Atlantis, which lay beyond the Pillars of Hercules (the entrance to the Strait of Gibraltar). It was a philosophical gambit as much as a geological or geographical proposition (science and philosophy really were fellow travelers), an imagining of a place way out there where power and pride and ambition held sway, a place eventually destroyed, as all such places must be when we waken from the dream. Or maybe it *was* a real place, destroyed by an earthquake or a volcano. He may have been inspired by accounts of Santorini, the Mediterranean island north of Crete that was devastated in about the seventeenth century BCE by one of the most powerful volcanic explosions we know of, darkening the sky, causing crop failure and famine throughout the Middle East and Egypt, and effectively changing the course of Mediterranean history. Or maybe it was a flood, for Plato said that Atlantis sank beneath the waves some twelve thousand years ago (when the last ice age was retreating and reshaping much of the world's water and land). Stories of the weird and wonderful transformations that took place during that period of climate change might have come down to him in story and song, and both the idea and the reality of Atlantis may have been Plato's image of a time when the waters rose and the land went under. Whatever the case, Atlantis has held people's imagination for nearly twenty-five hundred years, and roughly the same number of books have been written about it, establishing its location variously in the Mediterranean, off the west coast of Africa, in the Caribbean, and in the middle of the Pacific. Plato had certainly never seen Atlantis, and indeed the story he told was inside another story which was in turn inside another, like those boxes in the Haida Gwaii creation story. But neither have today's scientists ever seen the atoms they describe with such detail and delight, recounting stories that their instruments have given them, like the Egyptian priests whom Plato credited with the story of Atlantis. We should be careful not to dismiss the scientific imaginings of classical philosophers, for their storytelling styles are with us still.

So almost all early European and Islamic mappings of the world included images and icons of land surrounded by water. To the Romans, the surrounding or encircling sea was the River Oceanus. The Norse called it Uthal, or the Great Sea. To many sailors of ancient and medieval times it was the Green Sea of Gloom, with boiling waters and frozen wastes, monsters that defied description, and mysterious forces that didn't have a name. Sometimes there was a realm—an island—above and sometimes one below as well, often on an apparently flat earth; or, especially after the idea of a round earth took hold, the world was pictured as two hemispheres of land, one the mirror of the other where everything was backwards, including people's feet—which is why that place was called the antipodes. Some dismissed such a place as a philosophical joke or as a lie to mislead them from the truth, since they could not get their heads around the idea that folks on the other side of a round earth were upside down, with plants growing downwards and rain falling upwards. But Pliny the Elder in the first century CE had an answer, suggesting that "in regard to the problem of why those on the opposite side to us do not fall, we must ask in return whether those on the opposite side do not wonder that we do not fall."

It was also common for cartographers to place their particular home—Alexandria or Athens, Jerusalem or Mecca—right in the middle (just as mapmakers do today with their home turf). Thus centered, the place down under was referred to as the "austral" (from the Latin for south) or southern land. No one was sure whether it was a continent or an island; and indeed no one was sure of it at all. But its existence was presumed in geographical and philosophical traditions, and embraced by scholars who sought symmetry in a world with counterbalancing landmasses surrounded by ocean. Some Christian thinkers, intrigued by cataclysmic accounts, favored islands with strange or grotesque features instead of symmetrical landmasses, representing such singular islands as fragments of a whole, symbolic of a fallen and fractured world. But the idea, and slowly the reality, of the southern shores was hardwired into seafaring by the time of the European Renaissance and its expansion of trade and exploration on the high seas. One part of this great southern (is)land had an especially engaging name for awhile, a name that brings together ancient traditions of island travel with modern tourism. After Marco Polo had mentioned a kingdom he called Lucach, a printer's error resulted in it being identified on a map published in 1532 as Beach. Francis Drake, being a cavalier spirit, set out to find this exotic "Beach," apparently full of gold and elephants; but his cautious fellow captains persuaded him not to sail into what they thought was a gulf of one-way winds and currents.

In a world where about 70 percent of the surface is covered by water, beaches and shores are everywhere, forming borderlines between land and water. Botticelli's painting *The Birth of Venus* (ca. 1485) and Matthew Arnold's poem "Dover Beach" (1867) are only two of the thousands of representations of this kind of borderland, many of them conjuring up fear as well as fascination. In many places, waterfront real estate has become especially prized; shores of lakes and oceans have fetched prices far beyond reason. For ecologists, the edge or border is a place of peril as well as possibility; and since islands, in a very real sense, are all border, they become thresholds to a world of wonders in the stories and songs they sponsor.

Since ancient times, shores have also been the meeting point of different worlds and different peoples, sponsoring conflict as well as communion. Throughout history the sea has ensured that powers beyond human control hold sway on its shores; and even with all the hazards of the sea, the shore retains its own menacing authority. The most dangerous moments in an open boat—the kind that carried seafarers for thousands of years—were always launching from and returning to the shore, just as takeoff and landing still are for an airplane or a space shuttle.

There are two fundamental truths about humans and islands. The first is that until very recently, going to an island always meant journeying across a body of water, leaving behind the usual markers of meaning and value. The second truth is that we don't really understand what motivated people to do this, even though the history of island settlement has been a defining part of the human story from the day the first person left the land and ventured onto a river or lake or the sea on a quest for goodness or godliness or grub or gold. And islanders are not unanimous in their attitude toward the water that surrounds them, with tropical Pacific islanders viewing the sea around them as much friendlier than islanders in, say, the North Atlantic do—because their part of the ocean is in some ways indeed more "pacific" than the iceberg-clogged, storm-tossed far northern and southern seas, and because the great distances between islands in parts of the Pacific paradoxically seem to have created a sense not of island isolation but of ocean companionship, the water providing the currency of communication with other people.

Should islanders be considered colonists or castaways? The story of island habitation—how and when and why—is still controversial. The ability to fashion technologies for travel must figure in any answer, as perhaps does our instinct to cross boundaries, to make connections, to travel in between. Why we go, and why we stay, are among the most basic questions about our human occupation of this earth; which is why islands may be even more central to the human condition than language is, and why the history of island travel may define our deepest wants and needs (and not all of them admirable).

Islands clearly incorporate something fundamental about the human spirit. The stories and songs through which we make sense of the world represent both life as it is (or appears to be) and life as we wish it were or wonder whether it could be, both the so-called real world—its reality conditioned by our habits of thought and feeling—and the world of our imagination, shaped by our anxieties and desires. We try to keep these two worlds in balance, and to maintain some equilibrium between turning inward to ourselves and outward to the world. An island both illustrates and invites this kind of dual consciousness, which may be why some of our most enduring stories and myths have to do with islands. Faced with the difficulty of defining an island, perhaps we should take the advice of one of the wisest of ocean island scientists, Patrick Nunn, who proposes that islands are so completely built into our consciousness that we don't need a definition.

For a long time, many people have argued that we are most civilized, indeed most human, when we stop traveling and settle down. Others have seen something quintessentially human in our ability to dream about other places, design technologies to go there, and wander off. Islands are at the center of this human conundrum. To get to any island, you have to leave where you currently are and travel. On the other hand, islands are the perfect place to settle; once there, you cannot so easily go anywhere else. And there is something more, something that has to do with the human embrace of moments of wonder, of amazement, of awe. "I wish I were landing on her for the first time," said a seasoned Newfoundland fisherman as he approached a tiny island in the North Atlantic for the hundredth time, expressing a mixture of dread (for the landing was very dangerous) and delight (because thousands of birds were waiting to welcome him with a deafening chorus). Wonder is inevitably involved in island travel, no matter how routine. This wonder is circumscribed by wondering, as belief is surrounded by doubt, and islands by water.

Just as they were fundamental to ancient science and philosophy, islands have become central images in the modern social sciences, with concepts of the individual and society taking their cue from the psychology of islanded human beings and the sociology of communities as islands where interaction is unavoidable. Psychology and sociology have been joined by anthropology, economics, political science, and history in asking why humans travel to islands, why they stay there, and why some of them leave.

Eventually, such discussions return to a key set of questions. Is island living natural, something that happens in the normal course of human development? Or is it un-natural, prompted by particular circumstances? Are we island travelers by nature or by nurture? Are we pushed there by a crowded or cantankerous community—or are we pulled there by a desire for something different? Do independent people go to islands—or do people become independent once there? The questions go deep into our philosophical as well as historical consciousness, and echo ancient arguments about freedom and fate. And about foolishness. From early days, landlubbers have felt that anyone who gets in a boat and goes to sea for anything other than fish is either mad or bad or a bit of both. In medieval times in certain European, North African, and Middle Eastern communities, if you ventured far out to sea and managed to return, you would not be celebrated. You would lose your civil rights.

This suggests something unusual about those who find their way to islands, and make their home there. Maybe they are at the cutting edge of human evolution—or maybe they are cutting themselves off from the competitive challenges of progressive mainland life. Generations of anthropologists have had a fixation with this, scouring the islands of the world for evidence of either unique or universal human characteristics. One nineteenth-century anthropologist described the islands of the Pacific as museums or "cages in which their insulated occupants were shut in from external influence." "The sea selects and then protects her island folk," wrote the American geographer Ellen Churchill Semple early in the twentieth century, with a more positive spin. Of all geographical boundaries, she said, the most important is that between sea and land, and since most of the world is covered by water, "the human species bears a deeply ingrained insular character."

It seems we often need—or we want—to set out for lands across the water. "And then went down to the sea" is how Ezra Pound began his praise song to world literature, the *Cantos*. Similar lines can be found across much of the earth. The stories that Homer told opened that way, telling of "a wave-washed island rising at the center of the seas." "The first god was a gommier [gum tree]," writes Derek Walcott in his poem *Omeros* (the Greek word for Homer), referring to the tree used by islanders to make canoes for travel in the Caribbean. In the South Pacific during ancient times, ceremonies calling on the gods were performed on the completion of a canoe; in more recent times, the invocations are to Jesus and his fishermen friends.

The circumstances have varied, but the song itself has been much the same across the millennia. It was recited by Andrew Marvell in the 1600s, imagining a group of Puritans rowing to a fortunate isle they had heard about, the island of Bermuda—which had undergone a sea change in the century since it had first been named the Isle of Devils by the Spanish—and singing an elegant sea chanty to thank the God who "gave us this eternal spring / Which here enamels every thing, / And sends the fowls to us in care, / On daily visits through the air. / He hangs in shades the orange bright, /

Like golden lamps in a green night; / And does in the pomegranates close / Jewels more rich than Ormus shows. / He makes the figs our mouths to meet / And throws the melons at our feet." And to remind us that they went by boat, Marvell added: "And all the way, to guide their chime, / With falling oars they kept the time." Three centuries later another English islander, John Masefield, dreaming of "the gull's way and the whale's way," wrote the famous poem "Sea Fever" (1902): "I must go down to the seas again, to the lonely sea and the sky, / And all I ask is a tall ship and a star to steer her by."

Some seafarers have ventured over stormy seas and some through sheltered passages, but all of them have experienced the strangeness of sea voyages and, often enough, the sudden rightness of islands. The greatest migration in human history, when our early ancestors moved from Africa north and east through Asia, involved a major island expedition: as habitat and hunting capacities and climates changed and the exponential effects of population increase became overwhelming, some of them left the Asian mainland and traveled across the sea on rafts and dugouts to what is now New Guinea (the world's second largest island, after Greenland), which at that time—fifty or sixty thousand years ago—was part of the Australian landmass. Although we now call Australia a continent, it must have seemed like an island to the Aboriginal travelers who established one of humanity's earliest civilizations there.

Vagabondage—feeling "bound to go," in that wonderfully contradictory phrase—has probably been a part of life for people in all times and places. Ancient peoples would have heard stories about islands "out there," with a storyteller's assurance that if you traveled in a good way—whatever that might mean—you would sooner or later come across one of them; and setting out for one of the many islands on Earth would have been, in some circumstances, like going down the road in modern life. Rogue storms and outlaw escapades will have always played a role in fostering island travel, along with conflict at home and tales about life abroad. Extraordinary circumstances account for people taking astonishing risks, and we have lots of contemporary evidence for this, often with tragic results: people getting into rickety boats to seek refuge and a better life across the sea—from Cuba and Haiti to Florida; from northern Africa to southern Europe; and from Southeast Asia to Australia and North America.

At certain times, there may even have been a cultural bias *toward* going to sea, which over time would have become more like riding a horse or (in our day) driving a car or flying in an airplane. That may have been the case for the Arawak and Taino and Carib voyagers, and it was almost certainly so for the peoples who lived around the islands of Indonesia—there are seventeen thousand of these islands, with six thousand of them named and nearly a thousand now inhabited—where island hopping must have been bound into the human psyche, just as it was for peoples in the eastern Mediterranean where so much of Western literature had its beginnings. In East Africa, the name Swahili means "shore people"; and the various island chains in that part of the Indian Ocean framed by Africa and South Asia would have encouraged people to use the sea as a network of roads. "Whale roads" was the familiar phrase among ancient Scandinavians, who also traveled along archipelagoes, the sea connecting rather than dividing. People of the Atlantic, north and south, thought of themselves as shore people for millennia; and for

Inuit (Eskimo) peoples, the ice islands of the Arctic continue to be as much a part of their world as the sun and the moon and the stars. Even the Pacific Ocean, greater in size than all the land on Earth, was understood by its Polynesian peoples as a "sea of islands."

It is one thing to dream or talk about traveling to islands across the sea. Actually doing so is something else, no matter what the circumstances, and it represents one of humanity's most remarkable accomplishments and most extraordinary acts of faith. A covenant in wonder with the world. And a triumph of craft.

Chapter Two

Islands on the Horizon

Crossing the Waters

"Otaheite, or Ta[h]iti, an island in the Pacific Ocean [. . .] consists of two peninsulas, united by an isthmus about three miles in breadth. The greater of these is circular, and about twenty miles in diameter; and the latter about sixteen miles long and twelve broad. Both are surrounded by a reef of coral rocks, and the whole island is forty-four miles in circumference [. . .] The soil of the low maritime land, and of the valleys, is a rich blackish mould, remarkably fertile; but, in ascending the mountains, it changes into various veins of red, white, dark, yellow, and bluish earth. The stones exhibit everywhere the appearance of the action of fire; and the island has evidently had a volcanic origin [. . .] The more fertile spots, and even the mountainous districts, are covered with various useful vegetable productions, most of which grow spontaneously, and supply the natives with wholesome food. The most important of these are: the breadfruit tree, which seems peculiar to the Pacific Ocean, and which is found in the highest perfection at Otaheite; the coconut, which affords at once meat, drink, cloth, and oil; the plantain of various kinds; the chestnut, different in shape and size, but resembling that of Europe in taste; the evee, a yellow apple, a stone fruit, resembling a peach in flavour; yams, which grow wild in the mountains, from one to six feet in length; sweet potatoe, in great abundance, of an orange colour, resembling in taste the Jerusalem artichoke; tarro, a root from twelve to sixteen inches in length, and as much in girth, which is cultivated in wet soils, and the leaves of which are used like spinach; besides, a number of other roots and potatoes, made into pastes and puddings."

—*Edinburgh Encyclopaedia* (1830)

"TAHITI WAS CALLED Tahiti-nui, but first Havaiki by mistake, for our ancestor Maui [Polynesian trickster hero] [. . .] fished it up from the darkness of the deep ocean with the *kanehu* [bright, shining] fishhook which belonged to Tafai [a hero of ancient times]. The name of the hook was *Marotake* [to cause to be dry]. It was made from an *uhi* shell. Maui thought the land was the top of Fakarava Island [an atoll in the nearby Tuatomo archipelago], and as the name of Fakarava at that time was Havaiki, and it had lost its top from the anger of Pere [a localized version of Pele, the Hawaiian volcano goddess], Maui thought the land he had fished up was the top of Fakarava. So he called it Havaiki at first. But seeing it was a new land, a land not known before to men, a land not of one peak—as

Havaiki had been—but of many sharp points, he called it Tahiti-nui [from *hi*, to fish with hook and line]. He called it so because it was a new land, the one raised up by him, the one he fished up."

—From a story told by the Tuamotu islander Marerenui to J. L. Young, and written down in *The Journal of the Polynesian Society*, June 1898

Almost all islands share one thing: until the twentieth century, humans could only get to them by boat. And although leaving land and following a chain of islands in the Mediterranean or Caribbean or Indonesian seas was always an adventure, and seafarers from Europe and Africa and Asia and the Americas traveled to islands along their coastlines with remarkable skill and success for millennia, sailing on the open ocean was another story altogether. It still is; and sailing to the far islands in the Pacific has always been the ultimate test. Which leaves little doubt that Polynesians, who peopled those islands, were among the greatest ocean navigators in the history of the world.

The story of sea travel might not have begun with Pacific islanders, but it did take flight with them several thousand years ago, and their extraordinary seafaring has its counterpart in modern space travel. They needed to find their way to and from islands hundreds, even thousands, of miles apart and sometimes scarcely breaking the surface of the sea. And they needed craft to get back against the prevailing winds and currents that might carry them wide under sail. So the people of Polynesia (which is a modern term, meaning "many islands") had boats with remarkably stable design and construction, with sails that could catch the wind from different directions, and with the security that paddling provided. The particular ocean-going craft with which Polynesian seafaring is often associated—the outrigger canoe, where a separate float is attached to the main hull for stability—probably originated in South Asia. It seems to have been first used on the coasts of India and Sri Lanka, making its way eastward to the Torres Strait islands north of Australia and eventually throughout Polynesia (and also westward to Madagascar and the coast of East Africa). Two types were common: a small outrigger canoe, around thirty feet in length, used primarily for fishing or traveling short distances; and a larger vessel, either a double outrigger or two hulls connected by crossbeams (not unlike a modern catamaran), from fifty to a hundred feet long and capable of carrying a cargo of passengers and provisions sufficient for voyages of well over a month. The nautical technologies developed by the ancient Polynesian seafarers, still understood only in bits and pieces, allowed them to sail thousands of miles across the open ocean, even against the westerly currents and the east-to-west winds (generated by the rotation of the earth toward the east).

There would have been much local travel in the smaller craft, and the ability to launch and land boats safely would be learned by most members of any island community. Details of boat design, including wood type and sail rigging, varied across the region—which covers almost a third of the earth's surface and includes many thousands of islands—but construction with wood held together by fiber (to provide flexibility) was probably universal, allowing for give and take with big loads in heavy seas. The sails could swing about at different angles to the keel of the boat so that they would catch the wind coming from various directions (as distinct from sails fixed to the mast, more or less at a right angle to the keel, to catch the wind from behind). The canoes were sometimes leaky, so bailing out the water was crucial, and carefully carved wooden bailers were standard equipment. Many men and women would have participated in the makeup of these boats, bringing together the crafts of woodworking (for hull, mast, deck, and outrigger) and weaving (for sails and ropes). The Polynesian canoe, in all its many forms, ranks as one of the great triumphs of human technology.

It was exceptionally seaworthy, and Polynesian navigation was truly remarkable. The skills of open ocean navigation were probably mastered by only a few, as was also true for their European counterparts. But mastered they were, allowing the Polynesians to undertake long voyages deliberately, and successfully. They did not measure angles between stars and planets to determine latitude, as European navigators did (with instruments that developed from the astrolabe and quadrant into the modern sextant), but took direction from the passage of the sun during the day and the movement of the moon, the planets, the stars and their constellations at night. Some Polynesians also used a kind of wind compass, it is said, though its particulars seem to have been forgotten. But technologies aside, there was one fundamental difference between European and Polynesian navigation. For the Polynesian navigator, the boat was fixed while everything else was in motion, with the sun and the moon and stars as guides and a matrix of islands rather than the mainland providing points of reference, like a plotline. For the European navigator, on the other hand, the boat was moving, with everything else fixed at any given moment on a map or in relation to the sun or stars in the sky. Polynesian navigation identified "here" as where the boat was seen to be on the ocean, with reference points observed and observations coordinated on that ocean to bring traveler and destination—another island—together. In European navigation, "here" was where the boat was determined to be on the map, with reference points established on that map according to scheduled observations, or "fixes."

For the Polynesian navigator, then, even the ultimate point of reference—the island destination—moved through the stages of the voyage in relation to the boat, even though the navigator "knew" perfectly well that it did no such thing. Likewise a European navigator, using one of the various mapping "projections" that represented the curved surface of the earth on a plane surface, knew how to interpret the distortions that resulted. For example, in the Mercator projection (a map design invented by the Flemish cartographer Gerardus Mercator in the mid-sixteenth century) the straight lines are lines of compass bearing, a mapmaking illusion achieved by gradually increasing the distance between lines from the equator to the poles, which puts the size of land out of scale (with Australia seeming smaller than Greenland, when it is in fact three times as large). These are tricks of the trade. Sailors around the world work with them and live with the illusions they require. So do we all, in fact, imagining that we are living on a flat earth, standing right side up, and that the sun rises in the east and sets in the west.

Of course, there were many other signs that the Polynesian navigators relied upon, in addition to the night sky and the sun. They knew the currents, the wave configurations, and the prevailing winds in their region. They recognized land breezes and "sea markers," which were indicated not only by lines of seaweed and driftwood and the presence of types of fish, but also by the color of the water, especially near contrary currents. Precise directions, in shorter voyages, were given by birds returning to land every night after fishing, or taking off in the mornings; by sea turtles heading for shore; and by marine mammals such as dolphins on their way back from work or play. And smells would be noted, with breezes bringing the scent of green growth as early as a day before land was sighted. Skilled navigators would also feel and hear different wave configurations affecting the movement of the boat, and some of these would indicate the proximity of an island or the influence of a current. Reflections, such as the green of underwater atolls on the underside of clouds, would also help, along with the character of the clouds themselves. On longer voyages, migrating birds (such as the *kohoperoa*, or long-tailed cuckoo, which is well known on Samoa and Tonga and Raiatea and Tahiti and flies to New Zealand every October) could lead a Polynesian sailor to the island destination.

Every corner of the world has known island sailors, including Britain and Ireland and Iceland and the Mediterranean, as well as many South Asian and East Asian and African countries, and we sometimes score differences between them much too quickly. Seafarers everywhere share many of the same cognitive and cultural abilities—and an awareness of their fallibilities. Signs at sea are recognized by them all. James Cook, who learned much from Polynesian navigators during his three Pacific voyages (in the late 1760s and the 1770s), was successful because he brought his practices into line with their knowledge. Seagoing indicators similar to those the Polynesians relied on would have carried the Phoenicians across the

Mediterranean and the Vikings to Iceland and the Amerindians to Jamaica. What distinguished each of these seafaring traditions was *how* they used the universal signs at sea to set their course. For any navigator, it is never just a matter of noticing signs. Like trackers in the desert or readers in the library, they need to interpret these signs —and the ways of interpretation, like the scripts of different languages, vary widely and need to be learned in different locations and different cultures. Furthermore, navigational directions are always relative rather than absolute, like sounds in a word or words in a sentence; and as with languages, these relationships are specific to each situation. So navigators would need to remember the movements of the signs (the "words" of their watery worlds) and the relationships between them (their "grammar" and "syntax"). "All things are filled full of signs, and it is a wise man who can learn about one thing from another," said the Egyptian-Roman philosopher Plotinus in the third century CE, writing about what he called "the non-discursiveness of the intelligible world."

The Polynesians knew that the ocean was full of signs, and they knew how to interpret one thing from another at sea. To do this, their memories would have been finely tuned, often aided by mnemonic devices such as knotted strings. And stories and songs played a part in this remembering. Rua-nui, described as a "clever old Tahitian woman, then bent with age and eyes dim," in 1818 recited the following account of the birth of the heavenly bodies. It provided not only a genealogy but also a geography of the skies, and began: "Rua-tupua-nui (source-of-great-growth) was the origin; when he took to wife Atea-t'ao-nui (vast-expanse-of-great-bidding), there were born his princes, Shooting-stars; then followed the Moon; then followed the Sun; then followed the Comets; then followed Fa'a-iti (Little Valley / [i.e., the constellation] Perseus), Fa'a-nui (Great Valley / Auriga), and Fa'a-tapotupotu (Open Valley / Gemini), in King Clear-open-sky, which constellations are all in the North.

"Fa'a-nui (Auriga) dwelt with his wife Tahi-ari'i (Unique Sovereign / Capella in Auriga), and begat his prince Ta'urua (Great Festivity / Venus), who runs in the evening, and who heralds the night and the day, the stars, the moon, and the sun, as a compass to guide Hiro's ship at sea [Hiro was a Polynesian god who, like the Greek Hermes, specialized in trickery]. And there followed Ta'ero (Bacchus or Mercury), by the sun.

"Ta'urua (Great Venus) prepared his canoe, Mata-taui-noa (Continually-changing-face), and sailed along the west, to King South, and dwelt with his wife Rua-o-mere (cavern-of-parental-yearnings / Capricorn), the compass that stands on the southern side of the sky."

At night, when stars (and planets) were the most important indicator for a Polynesian sailor and were followed closely, each in turn would be replaced by another "guide star" when one rose too high or went below the horizon. Stars were like the songlines of the Aborigines in Australia. Not all stars, of course, were to be counted on. Some were known to be tricksters or troublemakers or just plain trivial; and contemporary navigators from the island of Anuta, between the Solomon and Fijian archipelagoes, still refer to stars in the major constellations as "carriers," while unnamed stars are called "common" or "foolish." Knowledge of the night sky was detailed, and laced together with lyric, narrative, and dramatic anecdotes of both natural and supernatural presences, with traits that many of us would recognize from the melodrama of Mediterranean and Scandinavian mythology.

A Samoan celestial catalog, for example, not only described red-faced Mata-memea (Mars) but also slow-goer Telengese (Sirius) and the balance-pole Amonga (Orion's Belt)—all of which showed the way for voyagers traveling from Samoa to Tonga. Cloudy weather or fog could of course interfere, but in most of the Central and South Pacific the visibility is exceptionally clear and cloud coverage limited to a few months. The early European explorers all confirmed this, and recent reports from the region suggest that clear skies can be expected at least two thirds of the year, with certain stars visible almost every night.

So even without the ability to determine east or west longitude (a disability shared with Europeans until the eighteenth century), Polynesians had techniques as trustworthy as those of the Europeans for determining direction. Also, exact navigation was not always necessary when islands were indicated by the character of currents and clouds, the movement of fish and birds, the sight and smell of drifting plants and leaves, and the keen eyes of seafarers (especially useful just before sunrise and just after sunset, when land is most easily spotted). And indeed, the same approximation was standard for European sailors during the seventeenth and eighteenth centuries, when both nautical charts and navigation involved estimates, and when no sailor relying on them could ever be sure precisely where he was, much less where the island he was looking for might be.

All navigators are storytellers as well as wayfinders, reading and interpreting both natural and man-made signs; and knowledge of these signs, passed down from generation to generation, has always been a crucial part of a navigational heritage. The principle is the same everywhere, though the medium may differ. Seafarers have relied on such songs and stories to tell them where—and sometimes why—to go and how to get there, taking them along on their voyages in memory or aided by drawings or knotted strings or manuscripts, or by ship's logs and sailor's journals. The best guide for navigation worldwide has often been the latest story told by those who have traveled that way before—after all, if they hadn't found a way, they wouldn't be telling the story.

Polynesians made their way by the quality of their attention to their stories as well as to the sea; and that attention had to be exquisitely focused. Order and relationship are everything at sea, and songs and stories guided them through as surely as the stars.

A traditional chant from Raiatea, near Tahiti, shows the detailed geographical knowledge of people in the South Pacific; it speaks of islands that are part of the Society Islands, the Tuamotu Archipelago, and the Marquesas Islands, as well as Hawaii. Such a song would have provided navigational direction as surely as a European nautical chart.

> Let more land grow from Havai[k]i! [Often identified with Raiatea, an island in French Polynesia and the legendary birthplace of the Polynesian people.] Spica is the star, and Aeuere is the king of Havai[k]i, the birthplace of lands.
>
> The morning Apparition rides upon the flying vapour, that rises from the chilly moisture.
>
> Bear thou on! Bear on and strike where? Strike upon the Sea-of-rank-odour in the borders of the west!

The sea casts up Vavau (Borabora), the first-born, with the fleet that consumes both ways, and Tupai, islets of the King.

Strike on! The sea casts up Maupiti, again it casts up Maupihaa, Scilly Island [Manuae] and Bellinghausen (Motuiti).

Bear thou on! Bear on and strike where? Strike east! The sea casts up Huahine of the fleet that adheres to the Master, in the sea of Marama.

Bear thou on and strike north! The sea casts up little Maiao of the birds in the sea of Marama.

Bear thou on! Bear on and strike where? The star Spica flies south, strike north-east!

The sea casts up Long-fleet in the rising waves of the Shaven-sea—the Shoal-of-Atolls (Paumotu) [Tuamotu Archipelago].

Bear thou on! Bear on and strike where? The vapour flies to the outer border of the Shaven-sea, strike there?

The sea casts up Honden Island, strike far north! The sea casts up the distant Fleet-of-clans (Marquesas) of the waves that rise up into towering billows! [. . .]

The sea of the Sooty Tern casts up the Island Cleared-by-the-heat-of-Heaven. There is cast up again the People's Head-land. [. . .]

Bear thou on! [. . .] Redness will grow, it will grow on the figurehead of the mountain at thine approach, as the sea ends over there!

> Angry flames shoot forth, redness grows, it grows upon the figurehead, as the sea ends over there.
>
> That is Aihi [the Hawaiian Islands], land of the great fishhook, land where the raging fire ever kindles, land drawn up through the undulation of the towering waves from the Foundation! Beyond is Oahu.

The first people to settle on what we now call Tahiti, about fifteen hundred years ago, were originally from the mainland of East Asia and Southeast Asia—though seafaring in the Pacific had flourished for thousands of years already, accelerated by cycles of climate change that caused the sea levels to rise and fall and dislocated coastal peoples. Some of them took to the ocean in search of new lands—islands—to call home, and over time they settled the atolls and islands and archipelagoes of Melanesia, Micronesia, and Polynesia. In the words of one ethnographer writing early in the twentieth century, these Pacific seafarers were "the champion explorer[s] of unknown seas of Neolithic times. For [. . .] long centuries the Asiatic tethered his ships to his continent ere he gained courage to take advantage of the six months' steady wind across the Indian Ocean; the Carthaginian crept cautiously down the West African coasts, tying his vessel to a tree each night lest he should go to sleep and lose her; your European got nervous when the coastline became dim, and Columbus felt his way across the Western Ocean while his half-crazed crew whined to their gods to keep them from falling over the edge of the world."

The sheer size of the Pacific made seafaring a special challenge. Winds and currents, which complicate all ocean travel, become major obstacles over such long distances if one doesn't have the knowledge or the technology to sail *against* these natural forces. European sailors, until late in the day, had neither, and so they found it hard to credit the Polynesians with the nautical and navigational expertise that they themselves lacked—but which the Polynesians would have needed to sail across the open ocean and colonize the Pacific islands. Instead, beginning in the eighteenth century, the Europeans made up a story that this colonization never happened, and that the supposed Pacific seafarers were really mainlanders who had retreated to the mountaintops when, once upon a time, the waters rose. Encouraged by the geological and fossil evidence of land bridges that once joined South America, Africa, and India, and by a religious conviction that land connections were the only way to account for the distribution of humans from a common origin, Europeans conjured up "lost continents," long ago sunken: Mu and Lemuria were two of them (still favored today in some New Age circles). These apologists for European shortcomings cast Polynesians as wild remnants of an imaginary ancient Asian civilization, or perhaps a lost tribe of Jews, or Athenian Greek voyagers—anything but the actual people who developed a nautical culture without peer in the history of the world, and a political culture that withstood the perils of isolation as well as any we know.

And just to be sure that the Polynesians were not credited with navigating the open ocean, the Europeans offered yet another theory to account for the settlement of the Pacific, based on a map from the mid-seventeenth century by the cartographer Arnold Colom that showed a string of islands sweeping southeast from New Guinea toward Cape Horn. The idea behind it was that only by means of such closely connected stepping-stone islands could the Polynesians have reached as far as they did. And then, after the map had turned out to be pure fiction, Thor Heyerdahl came along in the late 1940s with his *Kon-Tiki* raft expedition (and the extraordinarily popular book he wrote about it) to demonstrate that the Polynesian islands had been settled by *westward*-drifting South Americans. But linguistic, botanical, and cultural evidence was so unfriendly to his theory—as it has been to that of sunken continents—that it has now been thoroughly discounted.

The most important evidence of the migration of peoples throughout the Pacific islands comes in the form of oral histories and the languages in which they are told, which describe many of the voyages in remarkable detail and display similar linguistic and literary features throughout the region. There is also ancient pottery, in particular a type called Lapita, named for a beach on the west coast of New Caledonia where open-fired vessels, often with red-banded decoration, were found in the 1950s—and seen to be similar to pieces found thirty years earlier on Tonga, nearly fifteen hundred miles to the east. Ancient Lapita pottery was also identified on New Guinea and Fiji and most islands in between, confirming the existence of a civilization that had spread east into the Pacific over several millennia, peopling the islands and establishing traditions of story and song, of oceangoing craft and navigation, of music and dance and crafts, of food production and house construction and the harvesting of resources from the sea—traditions that are now seen as definitively Polynesian. The complex pattern of social relations between Pacific islanders, more communal than in the Caribbean, was illuminated early in the twentieth century by one of the greatest European island anthropologists, Bronislaw Malinowski, and described in his book *Argonauts of the Western Pacific* (1922). From the Trobriand Islands (east of New Guinea), where he lived for several years, he identified trade routes that were used to exchange not only commercial commodities but also items of little intrinsic, but of substantial symbolic, value—specifically necklaces and armbands—in a maritime cycle of gift-giving (called the Kula ring) that included islanders spread widely throughout the archipelagoes of the region. Such ceremonial traditions are part of both the ancient and the modern history of the Pacific islands, and they are emblematic of the larger ceremony of belief celebrated by getting into a boat (Malinowski described sailing in Trobriand canoes as floating, as if by a miracle) and heading out to sea. That is what the argonauts of the various Pacific isles have in common with each other—and what they share with island seafarers all over the world, including the Phoenician traders who sailed from the Middle East throughout the Mediterranean and beyond thousands of years ago, perhaps even reaching out to the archipelagoes of the Atlantic.

We are now sure that the Polynesians did indeed sail (rather than step from one disappearing mainland mountaintop to another) across the vast Pacific to its innumerable islands; and although they may have reached some by accident, evidence from the plants and animals they carried with them, as well as from the stories they told and the songs they sung, clearly shows that there were many deliberate island journeys. Their motives, as well as some of their methods, are still not fully understood—but neither are many of humanity's great adventures and achievements. Some Pacific islanders imagined the horizon as the eaves of a house, and if you went beyond them, you would find new dwelling places. Believing that theirs was a sea of islands, they *expected* to find more islands out there. Chance, but also their cosmologies and creation stories, would have played a part in an adventure—the game of their island-hopping lives—that had been carefully assessed and consciously joined. (As an analogy, we know from the analysis of oil exploration that once a general geological arena of hydrocarbons has been identified, random drilling is just as effective as targeted exploration.) The Polynesians would have recognized the same about their islanded ocean. What one chronicler describes as their fundamentally optimistic attitude toward ocean travel, and their confidence in their boat-building and navigational technology, would have made social contact across the sea as much a part of their lives as a trip across the city or to the neighboring town is for many of us. Of course there would occasionally have been other factors. Exile for the violation of a local taboo, for example, may have put some Polynesians to sea, just as it happened to the early British outlaws who were transported to Australia.

The seafaring peoples who populated the Pacific islands traveled distances on the open ocean that surprise even seasoned sailors today. Around 1300 BCE, they reached Fiji, Tonga, and Samoa, well over two thousand miles away from the islands off eastern New Guinea where their ancestors had settled, developing new spiritual and material relationships with Moana, the Great Ocean. Then, for reasons we don't yet understand, a millennium passed before there was settlement further east in the Pacific—though the appealing climate and geography of the Fijian, Tongan, and Samoan islands could be part of the reason, perhaps along with cycles of climate change (and there were several dramatic changes during this period) that altered wind patterns and sea levels. In the sixth century CE, Polynesians settled what we now know as the Cook Islands and French Polynesia, including Tahiti; in the seventh century, the Hawaiian Islands and Easter Island; and eventually, around the thirteenth century, New Zealand. About that time there came another round of climate change, this time apparently taking a more terrible toll. Nothing nourishes fear and loathing more surely than famine; and what we know from both the written records and the oral traditions of these islanders gives us a grim sense of the crisis that was created when the sea level fell by almost three feet, creating food shortages, severe conflicts, and malevolent cultural practices.

Tahiti, for its part, seems to have remained remarkably peaceable during this troubled time. When Europeans first arrived there in 1767, on the British survey frigate *Dolphin* commanded by Samuel Wallis, the ship's sailing master, George Robertson, described the island (in what would become an archetypal image) as "a great high mountain covered with clouds at the top." At first, they thought they might have found the legendary southern continent; but by the next morning, having sailed further around the shore, they realized it wasn't. As the fog lifted, they saw over a hundred canoes coming to greet them from the shore of this "pleasant and delightful island" (in Robertson's words), and Wallis decided to name it King George III Island. The British and the Tahitians traded supplies—nails and iron implements for hogs and chickens and fruit, which was a godsend to a crew suffering from scurvy, the scourge of seventeenth- and eighteenth-century ocean sailors (though the Dutch had known since the early 1600s that lemons, with their vitamin C, provided a cure). After some strategizing on the part of the Tahitian elders, who were anxious to make the best of this visit, the Europeans were received with generous hospitality, especially by the women whose beauty and candor delighted the sailors; and more goods were traded. The sailors also observed a solar eclipse, which allowed them to establish their longitude reasonably accurately; but it was disappointingly undramatic. (Celestial events could be very useful to European explorers, though their effect on indigenous peoples may have been misunderstood. On his last visit to Jamaica, in 1504, Columbus used the occasion of a spectacular lunar eclipse to promote the illusion of his direct line to the Almighty, and to bargain an increase in the trade of iron and implements for local food from the Taino. Or so he thought; for once again, and like many other travelers before and since, what he interpreted as primitive naiveté may have in fact been sophisticated negotiation.)

A year after Wallis, the French navigator Louis-Antoine de Bougainville arrived in Tahiti. Before that, Bougainville had been in Quebec with General Montcalm in his losing battle against the British under Major General Wolfe, part of a European imperial conflict that became known as the Seven Years War. It ended in 1763 with the Treaty of Paris, in which France traded a great part of its extensive territory in North America for the Caribbean islands of Guadeloupe and Martinique—at the time they thought they got the better of the deal. James Cook, who would sail to Tahiti a year after Bougainville, was on the British side in the Quebec battle on the Plains of Abraham; and if the coincidence seems surprising, it is a reminder that the history of the European exploration on the oceans of the world was written by comparatively few—a signal of just how rare and remarkable *their* achievement was. Polynesian navigation and colonization in the Pacific was extraordinary, to be sure; but we should also appreciate the accomplishments of those who sailed out from the European (and Asian) empires to the islands of the Pacific.

A few years after their defeat in Quebec in 1759, the French sent Bougainville to the Falkland archipelago to establish a colony. On his way south he visited Brazil, where his friend Philibert Commerçon, the naturalist on board his ship, discovered a lovely flowering plant—purple and pink and red and yellow and white and orange—to which he gave the name Bougainvillea. But when Bougainville reached the Falklands, he found out that the British were also starting a settlement; and its leader, John Byron, later governor of Newfoundland and grandfather of the great poet, claimed all the islands in the name of King George III. Enter the Spanish, reminding everybody of an edict from Pope Alexander VI, who in 1493 had drawn a line down the middle of the Atlantic on a map that included the still mostly unknown "new world," arbitrarily allocating everything east of the line to Portugal, and everything west of it to Spain. Nobody realized it at the time, but that gave Brazil to Portugal and the rest of the Americas—including the Falklands—to Spain. (The pope had been hoping to avoid disputes about island "ownership," but such disputes seem to be inevitable, and universal.) With God on their side and mammon at their disposal, the Spanish managed to buy off Bougainville in the Falklands with a substantial amount of money, and then they sailed against the British settlers, who pulled up stakes and left. Outraged, the British government threatened the Spanish, who withdrew—and the British settlers returned. Musical chairs on the islands of the world. An old story.

After leaving the Falklands, Bougainville continued on around Cape Horn to the Pacific. Although a year behind the British in reaching Tahiti, he laid the groundwork for French diplomatic success in the region now called French Polynesia, and his account of Tahiti has become a classic. Approaching the island, he noted that "the aspect of the coast offered us the most enchanting prospect. Notwithstanding the height of the mountains, they had no appearance of barrenness; every part was covered with woods. We could hardly believe our eyes when we saw a peak covered with trees right up to its summit [. . .] The less elevated lands were interspersed with meadows and copses; and along the coast ran a strip of low and level land covered with fields, bordering on one side the sea and on the other side the mountains."

Bougainville wrote only a few pages about Tahiti itself (published in 1771 in his book *A Voyage Round the World*), but they provided the textbook tale of tropical Pacific islands and the backstory for Tahiti's later reputation. "As we walked over the grass dotted here and there with fine fruit trees and intersected by little streams, I thought I had been transported to paradise. Everywhere we went, we found hospitality, peace, innocent joy and every appearance of happiness." He left his Edenic island after just ten days, mostly because his ships were in a difficult anchorage and in danger of being driven onto the beach by the winds. And he renamed the island La Nouvelle Cythère —the New Venus—after the island in the Mediterranean near where the goddess Venus was supposed to have risen from the sea. He also noted that it was known as *Taiti* by the natives.

Bougainville was a trained and not always tender-eyed observer, an experienced ocean navigator for whom accuracy was a condition of survival. So descriptions such as his should be taken seriously, even by those who distrust the accounts of early European explorers. To be sure, his words followed the literary and painterly conventions of the day, which irritate some people nowadays; but they were neither more nor less contrived than those of anyone else back then—not to mention our own travel writing and filmmaking. Spontaneity and sincerity have always been rhetorical conventions, and they change from time to time and from place to place, shaping the ways we tell the truth. The practiced eloquence of a naval officer such as Bougainville was also typical, for at the time military officers in European countries routinely received training in landscape painting and descriptive writing as part of their education. Other diary entries made by Bougainville on his voyage, about other encounters, confirm that he could be a scathing chronicler, so we have reason to accept his testimony about Tahiti. And Bougainville's eighteenth century was an age of enthusiasm, which we often distrust for no other reason than that it sometimes shows little trace of the irony that has become one of our modern literary conventions. Some years later, in 1793, José Bustamante, sailing under Spanish colors with the explorer Alejandro Malaspina, described the island of Vava'u in the Tonga archipelago in words that echo those of Bougainville. "Nothing can compare to the beautiful variety of scenery that met our gaze [. . .] The graceful harmony of the landscape and the confusion of evergreen trees scattered with flowers all spread before us the marvels of nature in her highest colors. In these delightful places the dullest imagination could not resist the sweet and peaceful sensations that they inspire. Here our minds were gently drawn to philosophical reflections on the life and happiness of these peoples [. . .] their tranquil existence in the midst of abundance and pleasure."

Tahiti is one of over a hundred islands and atolls in the five archipelagoes that make up French Polynesia: the Society Islands (including Tahiti, neighboring Moorea, Raiatea, and Bora Bora), the Marquesas Islands, the Austral Islands, the Gambier Islands, and the Tuamotu Archipelago. Tahiti's tropical conditions have made it a welcome home to many birds, to an abundance of marine animals, and to a wide range of plants, many of them having arrived with the cycles of human settlement. Its trees, the signature of so many islands, include coconut and mape (the chestnut of the Pacific); badamier (a tropical almond) and breadfruit; pandanus (also called screw pine) and mango (introduced in the nineteenth century); vi (or evee, a peach-flavored apple) and ay-hay (known in the Caribbean, to which it was transported, as pomme cythere or otaheite apple); noni (with medicinal properties prized by the Polynesians and widely popular today) and various bananas, with plantains being especially favored for porridge. Limes and lemons and tamarind and many other trees were brought by Europeans, but the original Polynesian settlers also established a number of the plants and trees on Tahiti. The tiare (of the gardenia family) is one of the species brought by the earliest settlers; it has small, white, fragrant flowers and is celebrated as

Tahiti's national flower. Other plants such as hibiscus and heliconia, anthurium and frangipani, along with roses and orchids, bring an international heritage to the island's many flowering plants, which are used by Tahitians to honor both the living and the dead.

As with almost every island, Tahiti is a sometime home for various seabirds that roam the oceans of the world, including albatrosses and petrels, boobies and frigate birds, red-billed and white-tailed tropic-birds, herons and kingfishers and gulls. Because it is so far from any continent—nearly four thousand miles from either Australia or North America, and almost five thousand miles from South America—the relatively few land birds on Tahiti, such as fruit pigeons, honeyeaters, weaver finches, cuckoo-shrikes, and kingfishers, would have come from New Guinea, probably stopping for rest (and recreation) on the Bismarck Archipelago, the Solomon Islands, Vanuatu, Fiji, and Samoa along the way. Other species, such as ducks and geese and turkeys, were bought by Europeans in the eighteenth century. There are few indigenous mammals, and even reptiles must have been put off by those distances. Some, such as the small blue-tailed skink (a lizard) could have traveled to Tahiti on driftwood; and people tell stories of lizards and even frogs, as well as spiders and beetles, falling from the sky after violent storms. Tahiti can remind us of many things—our supposedly Edenic past among them—but perhaps most of all that having wings, or excellent sailing skills, can make all the difference if you want to get to a remote island.

Fins and flippers would also work, of course, for the waters around Tahiti are swarming with fish and other sea creatures, including sharks and stingrays and mantas and barracudas, congers and snakes and moray eels, small anchovies and large albacore tuna, squid and cuttlefish (not a fish but a relative of the octopus), crabs and clams and oysters and beche-de-mer (also known as sea cucumbers), and a host of tropical fish: the beautiful angelfish and damselfish; the strange frogfish and triggerfish; the curiously named soldierfish and surgeonfish, hawkfish and parrotfish; the venomous turkeyfish and scorpionfish; along with groupers and snappers and puffers and grunts and gobies.

With this great wealth of seafood, Tahitians developed a narrative of advice for fishermen; a litany for the thirty nights of the moon is described as "a record for fishermen, recounting the nights when the fish run, this kind and that, and the seasons when these fish run, this month and that, and the days which are favorable for planting food plants." So, on the first night of the new moon (called Tireo), the storyteller recounts how "the fish have risen, all species of fish; the *iihi* have commenced to run on this night; the method of fishing is with a net, but the opening of the trap pocket should face the shallows," while a couple of nights later, "creatures of the sea propelled by their tails, those having hard shells, and also those that crawl, move about on this night; the method of fishing is with hook and line, and also by torch light in the deep sea for the *pao'e*, the *orare*, the *ma'una'una*, the *va'u*, and the *omuri*." On the tenth night, "the eyes of the fish are closed; they are asleep; this is a very fishless night, do not go fishing on this night lest you meet with disaster from fatigue"; but on the fourteenth night, "the *maito*, the *oturi*, and *para'i* and all kinds of red fish run at break of day; a net is the method of fishing for these fish; the following morning is a good time for planting food plants." And on the fifteenth night, "the moon has increased and reached her full development; all kinds of fish run on this night; a net is the fishing method, but the opening of the trap pocket should face towards the deep sea. This is the night when the sea crabs become soft-shelled."

Bougainville's account of Tahiti may have been the description of the tropical Pacific islands that came into popular consciousness—but it was Wallis who had set the stage for James Cook and the most important voyages of European scientific discovery in the Pacific until Charles Darwin sailed with Robert FitzRoy on HMS *Beagle* in the 1830s. Cook brought European and Polynesian science into conversation—and their religious heritages into a conflict that led to his death, on Hawaii, in 1779. His first voyage to the Pacific had several scientific ambitions, the first of which was to observe the so-called transit of Venus (i.e., the path of Venus across the sun) on Tahiti. It had been recognized for some time that the speed of the planet in transit, because of the earth's rotation, would vary according to where it was observed; and it was proposed that if the transit could be accurately timed from several carefully chosen locations around the globe, the differences could be used to calculate the distance from the earth to the sun (and, some suggested, the size of the universe itself). Transits of Venus come in pairs, eight years apart, and in the eighteenth century they took place in 1761 and 1769; but in the year 1761, weather and warfare in the locations selected for observation, as well as the particular trajectory of the planet, had ruined the possibilities for accurate calculation. The solar eclipse that Wallis had witnessed on Tahiti had allowed him to identify the island as a favorable site; and Cook was commissioned to go to Tahiti to observe the transit. As commander of HMS *Endeavour*, he sailed from Plymouth, on the south coast of England, in late August 1768, writing in his log: "At 2 p.m. got under Sail and put to Sea, having on board 94 Persons."

After a voyage of almost a year—the time it took back then to travel great distances rivals what we now associate with space travel to faraway destinations—Cook reached Tahiti three weeks ahead of schedule. His navigational skills were extraordinary, honed over many years at sea and in Canada, where he had charted a passage and set buoys through a section of the St. Lawrence River thought until then to be impassable for warships, guiding the British fleet through one night and setting up the defeat of the French in 1759. For several years in the 1760s, Cook then mapped the jigsaw-shaped coastline of Newfoundland. With his three voyages to the Pacific, Cook would become the greatest island surveyor and ocean cartographer in European history. The discipline he established aboard ship also signaled a change in European seafaring practice, and was almost as precise as his surveying. It included a schedule of watches that allowed for plenty of sleep, a regimen of cleanliness both personal (a cold bath daily, and bedding and clothing aired out every three days) and throughout the ship, and a diet that significantly diminished the risk of scurvy. He did observe the transit of Venus, in early June 1769, though a phenomenon known as black drop—a dark shade around the planet—made accurate calculations impossible once again. But there was always something else to do; and one of Cook's companions, the young naturalist Joseph Banks, collected flora and fauna, describing a thousand new species of plants, five hundred fish and as many birds, and also chronicled the cultural practices of the men and women of Tahiti, beginning the European anthropological as well as botanical fascination with island life. They spent three months there, the first Europeans to stay long enough on a Polynesian island to get to know something of the culture. Cook called the island O'Taheite, picking up its indigenous name, and named the archipelago the Society Islands, after the Royal Society that had sponsored his trip. Banks, whose interests were many—among other initiatives, he conducted ornithological and archaeological inquiries in Newfoundland and introduced nutmeg to Grenada—later became the Society's president.

Leaving Tahiti in July, Cook was off on his great geographic career, charting in his three voyages (1768–71, 1772–75, and 1776–79) the islands of the southern Pacific, surveying part of New Zealand (which had first been visited by the Dutch navigator Abel Tasman nearly one hundred thirty years earlier), as well as establishing the existence of the continent of Australia (and, by having listened to Maori elders, also confirming the *non*-existence of "Terra Australis Incognita," dreamed up by Europeans as a southern counterpart to the large northern continental masses). Cook landed in Australia's Botany Bay on April 29, 1770, first naming it Stingray Harbour and later changing the name to celebrate the exotic plants that Banks had collected there (their number increased by 10 percent the plant species catalogued by Carl Linnaeus earlier in the century); and he barely escaped disaster when he grounded on the Great Barrier Reef, saving the *Endeavour* only by throwing everything possible overboard and drawing a sail with wool and oakum under the ship to plug its leaks. They made it to shore, repaired the ship, and were on their way within two months.

Like all good navigators, Cook depended upon the knowledge of those who had sailed before him—not only the European but also the Asian seafarers, both of whom had been sailing the oceans of the world for a thousand years, and of course the Polynesians, who guided him and others who followed. There were material ambitions, of course; but they were often overtaken by the excitement of new knowledge. With Cook and his European contemporaries, seafaring took up surveying as a way of mapping the world's oceans and islands. His death on the island of Hawaii, in February 1779—caught up in a confusion of the spirits and the sovereigns of place—signaled the end of an extraordinary career; but it marked the beginning of a new era in the study of islands. When his ships arrived back in England after his last voyage, the plant and animal specimens they brought confirmed the importance of the organized collecting of species from the islands of the Pacific that has continued ever since.

Cook's journeys also nourished a new fascination with an old question: is paradise a place that knowledge inevitably corrupts? Certainly, the mutual acquisition of knowledge in these Pacific paradises conjured up a corresponding sense of the loss of innocence, just as it had with Adam and Eve in the first paradise. "Death is the mother of beauty," wrote the American poet Wallace Stevens in a poem called "Sunday Morning"—and the islands of the Pacific and Indian oceans were to become haunted by images both of beauty and of death, of paradises found and paradises lost. By the time Herman Melville reached the "paradise islands" of the Pacific in the 1840s, all that had enchanted the first European explorers seemed, to Melville, to be gone; and he quoted a Polynesian chant: "*A haree ta fow / A toro ta farraro / A now ta tarata* (The palm tree grows / The coral spreads / But man shall vanish)." Later visitors found images of love, but almost always inseparable from those of loss. When Paul

Gauguin wrote on his famous 1897 painting of a nude Tahitian girl lying on a bed the single word NEVERMORE, he conjured up the sad refrain of Edgar Allen Poe's laconic raven, which sits on a windowsill in Gauguin's painting.

But the Tahitian paradise did not die. It was kept alive in the everyday by many Tahitians, and in the imagination by writers such as Robert Louis Stevenson and Somerset Maugham; and its allure has been renewed recently by the enterprise of those who promote the appeal of island wonders and the delights of comfortable isolation (if only for a week or two). The European fascination with Tahiti, which began with the eighteenth-century praise of the "noble savage" sponsored by Jean-Jacques Rousseau, continues. The cycle of life and death and rebirth is the true history of islands, as geologists insist; and in this story, Tahiti and the islands of the Pacific have a central place.

Seafarers have always been ready for moments of awe, for while they are afraid of many things, they cannot be afraid of wonder. For a sailor, being ready for surprises is the key to survival. So wonder and wondering were the nouns and verbs of early exploration, and have been the staple of sailing stories from ancient to modern times. Of course, those eighteenth-century European navigators would also have had in mind entertaining an audience back home—what storytelling traveler does not—but they were at work in their logs and journals, which provided whale-road maps for those who followed them. Their lives, and the lives of those who came after, depended on their telling the truth about what they saw on shore, as well as at sea. Their contemporaries could read through the conventions, as we see through those of our own day; and although the kind of landfall described in accounts of Tahiti was a haven for sailors, it was never without its hazards.

Also, the spiritual dimension of these enthusiastic descriptions was not incidental, either to Tahiti or to the human history of islands. In the tropics, the imaginations of northern seafarers conjured up paradise, with Adam and Eve lounging or loitering by a tree. We may be tempted to dismiss this as idle romanticism, but the association was appropriate in more ways than we might expect. It was the search for knowledge as much as for spices or subsistence rations that brought these islands into the center of European consciousness; and paradise, after all, was the place where knowledge was first found in the story many of these sailors carried with them. Gardens (including the Garden of Eden) were routinely described in language normally applied to islands—and vice versa; and tropical islands had long had a reputation as sanctuaries, from which their sometime sensuality was never entirely separable. The contradictions continue to the present. Jamaica began its European history with a description of its beauty, and even after several bittersweet centuries it has maintained that image, its sensuous pleasures advertised to tourists around the world, its spiritual authority firmly established in the ceremonies of Rastafari and the songs of Bob Marley and his reggae apostles, and its complicated history of slavery and the seashore bound together in the lyrics of a song written by Irving Burgie and sung by Harry Belafonte: "This is my island in the sun / Where my people have toiled since time begun / I may sail on many a sea / Her shores will always be home to me."

The Pacific Ocean has long been a favorite site for sacred and paradisal spaces. In Dante's fourteenth-century Purgatorio, redemption was to be found on an island in the southern hemisphere on the opposite side of the world from Jerusalem, and Thomas More's sixteenth-century Utopia was set on an island somewhere in the Pacific. The Portuguese Pedro Fernández de Quirós, sailing (under Spanish colors) at the beginning of the seventeenth century in search of the southern landmass Terra Australis Incognita, had a geography of yet-to-be discovered islands ready at hand for describing his latest landfall. He did not have an easy voyage, with a mutinous crew on board and hostility on shore, and he failed to find anything of commercial value, much less the legendary Austral-ian landfall which was supposed to be a place of fabulous riches hinted at in Inca stories about South Pacific islands collected by the Spanish in Peru. But he turned his travels into a quest for a place of purity, and when he reached the islands of Vanuatu (formerly called the New Hebrides), he set out to establish a New Jerusalem and wrote of one of the islands there—called Espiritu Santo—that it was an "earthly paradise." It is said to have inspired two of James Michener's stories in his *Tales of the South Pacific* (1947), which became the basis for the musical and movie *South Pacific*.

For all his commercial ambitions, Columbus, too, sailed the seas in praise of his God, and he established his own spiritual as well as material covenant in wonder with the world he "discovered." So it is no surprise that his most memorable words were in the language of wonder, certainly not restricted to matters of trade and commerce. Like Marvell's Puritan travelers rowing toward Bermuda, and like Bougainville in the South Pacific a century later, Columbus was inspired by the beauty as well as the bounty of the islands he visited in the Caribbean, describing them in his journal as "very lovely and green and fertile . . . All is as green and the vegetation is as that of Andalusia in April . . . The singing of little birds is such that it seems that a man could never wish to leave this place; the flocks of parrots darken the sun, and there are large and small birds of so many different kinds and so unlike ours, that it is a marvel. There are, moreover, trees of a thousand types, all with their various fruits and all scented, so that it is a wonder."

Behind these comments lay another conviction, closely connected to principles that developed centuries later. These men believed, as psychologists and sociologists later proposed, that the environment was a determining factor in human development, nurturing cultural conditions, and that beautiful surroundings produce beautiful people. Simpleminded, perhaps; but the idea would later have a profound influence on everything from architecture to education. And in fact it is not so far removed from the theory of evolution.

Wonder was the key—and seafarers published wonderful accounts of their voyages. In seventeenth- and eighteenth-century Europe, writing about maritime travels was like writing political or business memoirs these days. Everybody did it—and many people actually read them. One of these seafaring chroniclers was William Dampier, a notable explorer and a prolific author, whose *A New Voyage Round the World* (1697) was a bestseller, reprinted five times in six years, followed by a half dozen further volumes describing his three circumnavigations of the globe and his adventures along the way. Several of his stories provided plots to the journalist Daniel Defoe, though the story of Alexander Selkirk—the inspiration for

Robinson Crusoe—came from a privateer named Woodes Rogers, who rescued Selkirk and wrote about his four-year sojourn on Más a Tierra in *A Cruising Voyage Round the World*, published in 1712. Dampier was in fact on that trip, as well as on the one when Selkirk was marooned; and his travels caught the attention of another of the great writers of that time, Jonathan Swift. "My cousin Dampier . . . ," begins Swift's fictional account of Lemuel Gulliver.

The success of the privateer chronicles makes even more sense when we realize that privateers were the entrepreneurs of what was then the "new" global economy—buccaneers with government backing, looking for loot and the next place to land. And some were very successful. Those who backed Francis Drake as he circumnavigated the world, for example, made 4,700 percent on their investment; and this was used by Queen Elizabeth (as no less an authority than John Maynard Keynes pointed out) both to clear the British foreign debt and to bankroll one of the major trading ventures of the time, the Levant Company. Robber barons had their start as robber buccaneers.

And the novel had its start as popular nonfiction, adapting fact with a flair that appealed to a wide range of readers. In the beginning—which is to say, in eighteenth-century Europe—novelists tried to blur the line by pretending that theirs was a "real" story, taken from actual documents in some unidentified treasure trove. And sometimes it was, in narrative poems as well as in prose. Scenes in Coleridge's "Rime of the Ancient Mariner," for instance, came from the journals of the privateer George Shelvocke, who published his *Voyage Round the World by Way of the Great South Sea* in 1723, writing about an incident where his second in command shot an albatross and his ship was later wrecked on the very same island, curiously enough, where Selkirk had been "exiled." These and scores of other books displayed the language of wonder and surprise in what had become the style of storytelling.

The connection of navigation with knowledge is preserved in the language we use when we "navigate" the World Wide Web with our "search engines." The American New Critic John Crowe Ransom once described a literary text as one that reveals "the kind of knowledge by which we must know what we have arranged that we shall not know otherwise." Navigation is like that; and its different traditions—Polynesian and European, for instance—remind us of the tricky relationship it negotiates between the interpretation of signs and the belief in them. The word "negotiate" derives from the Latin *otium*, which means ease; and good negotiators, like good navigators, are never at ease. Hermes, the messenger, or carrier, of signs in the pantheon of Greek gods was known as a trickster; and hermeneutics, the tricky craft of interpreting texts, is named after him. In what is sometimes called the hermeneutic circle, beloved by biblical scholars, there is no interpretation without belief and no belief without interpretation. Both belief and interpretation are always surrounded by doubt, and that is certainly the case with navigation at sea. It's a tricky business. The word for trick in Homer was *dolos*, and its first reference was to a fishhook, like the one used by the Polynesian trickster Maui to pull up islands.

Sailors are almost always unsure about things, and they rely on tricks of their trade to deal with the tricks that the sea plays on the unwary, which may be why they are renowned for their superstitions.

Good navigators sometimes say they are always almost lost. And indeed, until a few decades ago on European and American navigational charts, nobody was exactly sure where many of the ocean's reefs and rock outcrops were located. Some small islands hadn't been properly charted, some of them moved about with the shifting tides and currents, and some of them were known only by reputation. So a seafarer could never really know. Even as late as the 1950s, the nautical charts of the South Pacific were filled with islands marked P.D. (Position Doubtful) and E.D. (Existence Doubtful). This wasn't encouraging; but it did keep sailors watching, and wondering. Western navigators use storytelling language—the language of wonder—when they "plot" their position on a chart, and they do so in pencil, because such a plotline is always open to revision.

We know more about this state of mind than we may realize. The word "disoriented," which we routinely use to describe how we feel when we're not sure where (or how) we are, was first coined centuries ago to describe the experience of seafarers navigating far from their center of certainties. For medieval Christian and Jewish and Islamic sailors who went out from the Mediterranean onto the Atlantic Ocean, that center was in the East—the Orient—where they identified their spiritual (and often also their secular) home. If they ventured too far into the western sea—as they occasionally did—they became *dis*-oriented, alienated from their home. In the northern hemisphere, the constellation Ursa Major (the Great Bear) was a main marker, recognized for thousands of years under various names: a mother bear, a butcher's cleaver, a farmer's wagon, a drinking gourd, a big dipper. Polaris (the North Star) poured off the dipper, likewise guiding sailors and other travelers north of the equator.

(On land, too, the constellation was a guide; for example, the song "Follow the Drinking Gourd" was believed to map out the route north to Canada for those escaping slavery in the southern United States.) Portuguese sailors, sailing to the south, came up with their own word to describe how they felt when they were far from their home and the North Star by which they usually navigated. They said they were *desnoreados*. Dis-northered. *Desnoreados*, like disoriented, still is used as a synonym for—well, for being "at sea."

Before satellite technology gave us GPS, being out of sight of land in European navigation meant that nothing was certain, because everything was (relatively speaking) moving: the wind, the waves, the current, the tide, the stars, the horizon; it was a place where (at least when using European navigational techniques) sailors had to figure out—"fix"—their position at any given moment not only by the location of the stars and the movement of the currents and wind, but also by a method called Dead Reckoning, used since well before Columbus. It's not a cheerful phrase. Some say it is a misspelling of the abbreviation "ded." for reckoning by deduction, others that it refers to observation without the lively movement of the earth and the stars. Dead Reckoning involves estimating a ship's current position based on a previous position that the navigator has fixed on the chart, as well as by the ship's direction, speed, and distance since that fix—which of course is itself based on an estimation.

So while it was important to know the location of the islands and outcrops, it was also important to know where *your* ship was on the map; and of course all of it was interdependent, since the location of islands was established by sailors, not by satellites. The measurement of latitude had been reasonably well developed—at least in principle—

by the ancient Greeks, using the position of the stars and sun relative to the horizon. Instruments called astrolabes were in use by astronomers a couple of thousand years ago to measure the angle of stars above the horizon, and were improved over the centuries for use at sea, so latitude could be fairly accurately established by a combination of readings at sea and (when landfall was made) on shore. Establishing compass direction, a crucial navigational item for Europeans, began in the twelfth century with a magnetized iron needle on a wooden float in a bowl of water, refined into what we would recognize as a modern compass by the fourteenth century, and then improved over time by correcting for the difference both between true north and magnetic north and for variations in different locations.

Calculating longitude was a different matter altogether in European navigation. Since longitudinal lines are not parallel to each other (unlike the parallels of latitude), there were no natural points of reference; and it wasn't until the eighteenth century that sailors were able to calculate their longitude—and therefore their position—reasonably accurately with the help of chronometers. For earlier seafarers, the cost of not being able to do so could be very high. In 1707, there was a particularly deadly toll on a fleet of British naval vessels sailing by the Scilly Isles off Cornwall. In stormy weather, four ships were wrecked and fifteen hundred sailors drowned; and in short order the British parliament—an island parliament, of course—passed the Longitude Act, offering a prize of twenty thousand pound sterling for the first person to come up with an accurate method for calculating longitude at sea. Some thought the solution might be found in a different calculation of the movement of the stars; but the trick turned out to be timekeeping with instruments precise enough to compensate for the 15 degrees that the earth rotates every hour (i.e., 360 degrees in 24 hours). By 1735, the self-taught clockmaker John Harrison had developed a chronometer that would maintain its accuracy at sea—and sailing the seas became a lot safer thereafter.

But even with all their instruments, European sailors who charted the oceans could only do so inaccurately. They did their best, of course; but with determinations dependent on the accuracy of a series of earlier fixes and a rough calculation of speed and direction (taking into account wind and current), they were sometimes off by hundreds or even thousands of miles, especially if they had gone through a storm or two. There are stories of European explorers and adventurers—and probably Polynesians, too, for all their knowledge and skill—sailing right through entire archipelagoes without sighting a single island. The accuracy we now take for granted is very recent. It has only been in the last fifty years that precise positions have been determined, and almost all the maps redrawn. Accounts provided in authoritative atlases and encyclopedias up until the 1960s were often either apologetically vague or absolutely wrong.

The first word recorded in what became the English language was written down by a cleric named Gildas in a sixth-century CE sermon. It was *cyulis*, and is the root of our modern word "keel"; but it referred then—and in some seafaring communities still does—to a particular kind of flat-bottomed boat. These old English keels were generally clinker-built (also called lapstrake), with boards laid over each other so that they overlapped along their edges, a technique perfected by the Vikings for their longboats.

The distinction between boats and ships is still disputed and often ignored, though there is general agreement that a boat can be carried on a ship. Be that as it may, the archaeological evidence of early boat-building is limited, for material traces have mostly disintegrated and disappeared, washed away or buried under rising water. Also, people would have taken the wood from old boats to build houses; and what was left would have been burned, or drifted into oblivion. The search for the ark on Mount Ararat continues, but like all wooden boats it is unlikely to have lasted long (if in fact it ever existed). Images of boats have been found in caves and tombs around the world and going back at least ten thousand years, many of the earliest of these in Central and South Asia. Some of them commemorate journeys that only gods and the dead would undertake, while others celebrate human travel on the water. We can be sure that along the oceans' coastlines, underwater kelp forests would have nourished a wide variety of flora and fauna that were both gathered on the shore and hunted on the water by people in boats; barbed hooks and nets for fishing have been found in ancient sites, and seals and sea lions would have been relatively easy prey.

Indeed all over the world, for at least fifty thousand years—and probably much longer—people have been out on the water in rafts and dugout canoes, first paddling and then rowing and sailing. It was back then that Aboriginal peoples crossed the sea to reach the continent of Australia. Stone tools found recently on the island of Crete suggest that island travel in the Mediterranean began even earlier. These tools were embedded in terraces formed along the shore by the slow rise of the island, forced up by colliding continental plates at a rate of about an inch every twenty years, which dates them to roughly 130,000 years ago. Geologists confirm that Crete has been an island for millions of years; and since the rock used to make the tools is of Cretan origin, the toolmakers must have traveled to the island by boat—which opens up the possibility that prehuman cultures may have set out to sea.

We don't know exactly when and where the use of sails began, but there are images from seven thousand years ago on pottery in Mesopotamia, the land between two rivers, that are sometimes interpreted as sailboats and sometimes as bowls for spinning yarn. This is an interesting uncertainty, for it would have been with the spinning and weaving of textiles and ropes that sailing itself became possible. The timing would also coincide with the domestication of horses, which occurred around then on the steppes of Asia, and the connection between working horses and sailing boats is preserved in the language and the logic of each. Horse technologies—saddles and bridles and stirrups and bits—are called tack, close cousin to nautical tackle; and the word rigging is used around both horses and sailboats. More significantly, perhaps, riding a horse and sailing a boat both require a sort of indirection, working the angles, establishing a new relationship with the forces of the natural world—a relationship in which a balance between surrender and control is indispensable. It was with this balance, and this technology, that humans shaped their civilizations on the continents and islands of the world, riding on the steppes and sailing on the seas.

Knowledge spread quickly. The Egyptians were certainly sailing boats five or six thousand years ago, and soon the Mediterranean was a sea of sails. Square-rigged sails were probably the first to be in common use in that part of the world. (While they were usually rectangular, they are not called "square" sails because of their shape, but because the sail is hung in a right angle to the keel of the boat.) They were the standard for awhile, but there was one problem: they required steady wind from behind. "Beating to windward" (sailing at an angle into the wind), on the other hand, was difficult with square rigging alone. Before long, war and trade and the everyday imperatives of transportation by sea generated new technologies, and maritime peoples eventually came up with a sail that could take the wind on either side. In European accounts, it was credited to the Romans (and often called "lateen," from the Latin spoken around the Mediterranean), though it was popularized by Arab traders. The lateen sail was triangular, and the spar (pole)—to which one side of the sail was fastened—was attached high up the mast at an angle of about 45 degrees. One corner of the sail was fixed toward the bow, while the other end was held by a rope and free to catch the wind on either side. It produced much better results than a square sail against a headwind (although it was not as good in a tailwind), and it was favored for small boats used for short runs in various directions. To this day, the lateen is standard rigging on some recreational sailboats, and is still seen on fishing boats around the world.

The lateen sail and rigging became common in the Mediterranean and Indian oceans, but it had been developed elsewhere, too, and recognized for its efficiency. Ferdinand Magellan saw lateen sails on Polynesian craft among the Pacific islands that he reached in 1521 following his turn around South America; and his chronicler, Antonio Pigafetta, included a drawing of one on a map of the Mariana Islands. Chinese junks would have plied the China Sea and beyond for centuries with their own form of this sail and rigging, and the great Chinese navigator Zheng He sailed in the early fifteenth century across the Indian Ocean to Africa with a fleet of them. Balangays in the Philippines, with sails that also turned around the masts to allow the boat to sail into the wind, had been in use for over a thousand years when the Europeans first saw them.

The advantages of sailing over rowing were obvious, especially if the wind was blowing from behind strongly enough to overcome any tides or currents that might be running against the boat. But paddling or rowing—like its latter-day replacement, the motor—always provided an important alternative, especially where lateen sails were not yet common. In one famous ancient account, from about 500 BCE, the Carthaginian navigator Hanno sailed a fleet of galleys down the Moroccan coast and successfully returned, each galley rowed by fifty oarsmen. (The opinion of the rowers, pulling for days in blistering heat, was not recorded.) Winds and currents presented a particular challenge to early European sailors down the northwest coast of Africa—after they had left the relative safety of the Mediterranean Sea—with winds favorable for heading south but not for returning home, particularly between the Canary Islands and the mainland of Africa where there is a strong southerly current. And for a long time this coastal route was the only one that most sailors dared take, since the alternative would have carried them far off the coast into the open ocean where rowing was not an option and where the position of islands was still uncertain.

Landlubbers often make the mistake of assuming that the most dangerous hazards at sea are high winds and dense fogs and what

E. J. Pratt once called "the primal hungers of a reef." But *no* wind can also be deadly, especially for a ship relying solely on sail. Coleridge gave a grim account of this in his "Rime of the Ancient Mariner"; and the horse latitudes, just north and south of the equator near the Tropics of Cancer and Capricorn, were so named because when ships were becalmed there, sailors would sometimes throw cargo horses overboard in desperation to preserve water and lighten their load.

There is a poignant account in the journals of the Dutch explorer Jakob Le Maire of the situation facing travelers, long before the days of steam or gasoline, in sailing boats not built to be rowed or paddled. In 1616, Le Maire had found a new route to the Pacific round Cape Horn (named after one of his ships, the *Hoorn*), avoiding the dangerous straits that Ferdinand Magellan had earlier navigated. On his way north, Le Maire kept close to the coast of Chile so that he would be carried along by the northerly Humboldt Current and make landfall on the group of islands first discovered by Juan Fernández half a century earlier. According to the practice of the time, Fernández had seeded these islands with pigs and goats to provide food for ships that stopped by or ran aground. The Juan Fernández Islands eventually became a regular stop for sailors traveling up that coast. (One of them was where Alexander Selkirk would later jump ship.)

Le Maire sighted these islands one sunny day, running with a good southeasterly wind behind him; but when he took his ship up the west side of the main island, his wind was cut off by the mountains, leaving him at the mercy of the current—which carried his ship right on past the island. His account stands for hundreds, if not thousands, of similar (but mostly untold) stories. "Since we were unable to get close enough to the shore to anchor, we sent out our boat towards the land. In the evening she came back with news that the shore was fit for an anchorage and that the land was verdant and full of green trees and fresh water. The men had caught a great quantity of fish: no sooner had they dropped a hook in the water than a fish was upon it, so they had nothing to do but draw up fish without stopping. These tidings made the crew very glad, especially those who had scurvy. In the night the wind fell altogether, so that we drifted north with the current. [On the next day] we were again close to the land, but try as we might we could not get close enough to anchor. We sent the boat ashore, and whilst some men were getting water the others caught close on two ton of fish. They then had to leave this fine island, without enjoying it further. [The following day] we had drifted about four miles past the island, notwithstanding that we had for 48 hours done our best to reach it under sail, which vexed us exceedingly, seeing it was impossible to make it. It was therefore agreed that we should leave the islands and pursue our voyage—to the very great sorrow of the sick who saw all their hopes lost." And this from one of the great sailors of his day, who on his way to the Pacific had stopped at Sierra Leone and bargained for thousands of lemons to protect his crew against scurvy, a century and a half before it became common sense.

Islands sometimes offer other challenges to those who come by, often signaled by local names. One island on the North Pacific coast was named Xwlil'xhwm (Fast-Drumming Ground) in the language of the Squamish peoples who have lived there for thousands of years, the name catching the sound of the waves driven by currents that swirl around the capes and coves and winds sweeping down from the mountains against their great cedar dugout canoes; and a channel where the waters troubled the rocks and where the currents swirled under the swell of the sea was called Kwum'shnam (Thumping Feet). Strong currents surround the Faroe Islands in the North Atlantic, and the largest island, Streymoy, is called the Isle of Currents. Old accounts speak of how the Faroese fishing boats were once powered by eight or more oarsmen, rather than by sail, because unpredictable whirling winds were common under the mountains—just as they are by the headland called Blow Me Down on Newfoundland. Islands all over the world have their individual character, of course, with nautical and navigational strategies worked out by their residents over generations—particular to each place, to the technology of the time, and (not insignificantly) to the culture of the people. The navigational skill of sailors in such waters is legendary—it was said that the direction taken by a louse on the thwarts could show the Faroese sailors their course—though as with the Polynesians, such signs are sometimes discounted by others, since few outside a small circle can understand them.

While sustenance and settlement seem to have been the main reasons the early Polynesians traveled to the far islands of the Pacific, it was science and the harvesting of sea mammals that took Europeans there. But first and foremost it was spices. Trade in spices began several thousand years ago but picked up pace in the later Middle Ages, when the rise of the Ottoman Empire, with its intimidating travel regulations, made overland routes between Europe and the

Far East difficult. Spices had gained social status as a luxury item by that time, at least in the Middle East and Europe, and not only cooking ingredients but incense of various kinds, some medicines, certain gems, and copper and tin were routinely included within the category. Some spices also gained a reputation as aphrodisiacs, and a few as agents of transcendental insight. For a long time, the most valuable spices of any description came from the East: pepper and ginger from India, cinnamon from Sri Lanka (then known as Ceylon), and nutmeg and cloves from the archipelago south of the Philippines called the Moluccas. These were the so-called Spice Islands, though other islands, such as Zanzibar and Mafia Island (both off the coast of Tanzania), also took the name. But it was the Moluccas that became both the goal of European sea-traders in the fifteenth, sixteenth, and seventeenth centuries and the heart of an overseas Dutch empire that for a time exercised control over a profitable trade in camphor and quinine and various dyes as well as spices. Though their monopoly was eventually broken when the British planted nutmeg in Sri Lanka, it was in the search for a route to the East Indian and Indonesian spices that the islands of the West Indies were "discovered" by Europeans. Christopher Columbus and John Cabot were among the first to sail west from Europe in the hopes of reaching those fabled Spice Islands, Columbus taking the southerly route across the Atlantic while Cabot—an Italian who had been in the spice trade in the Mediterranean but sailed in 1497 under British sponsorship—looked for a northwest passage to India. Columbus and Cabot were followed in their search for a route by the likes of Ferdinand Magellan and Jakob Le Maire.

When Columbus bumped into the islands of the Caribbean, it was little wonder that he thought he had made it to the Far East, for the armchair geographers of his day had underestimated the circumference of the world by nearly a third. For millennia, the greatest mathematicians of the Mediterranean had been trying to calculate the circumference, measuring the distance between two fixed points and calculating the difference in the altitude of the sun between them. In ancient times, one of these points was usually Alexandria, a center of mathematical inquiry; and a remarkably accurate calculation by the geographer Eratosthenes in the third century BCE was based on measurements taken there and at Aswan, the distance between them being precisely measurable overland. Another measurement of the earth's circumference, by Posidonius a century later, was based on the distance between Alexandria and the island of Rhodes, which were further apart and therefore likely to provide a more accurate calculation of the difference in the sun's altitude. But Rhodes being an island, the distance across the water between it and Alexandria was much more difficult to measure. Accordingly, Posidonius's calculation of the circumference of the earth was out by nearly 30 percent. However, the great geographer Ptolemy accepted it in the second century CE, and it was written into the record on the charts that most sailors used for the next thousand years. Marco Polo recognized that Ptolemy was wrong in his overland distances, but unfortunately—knowing something about land travel but much less about seafaring—he compensated for this by narrowing the oceans. So Columbus went to sea with Marco Polo in his pocket, visions of spices in his head, and mistaken figures for the circumference of the world on his charts.

But there were those who dreamed of a different route from Europe to the spice and textile markets of the East. In the early fifteenth century, before Columbus was even born, the Portuguese prince Henry—who liked to be called "the Navigator"—was sending explorers out into the Atlantic and down the western coast of Africa. Portuguese navigators had been sailing into uncharted waters for a long time, and in their travels they came across various islands, including the Canary Islands and Madeira, the Azores and the Cape Verde Islands (and they were almost certainly also working the rich fishing grounds around Newfoundland).

Henry had made his fortune with a commercial monopoly on soap (and probably also with occasional piracy), and despite some spiritual flourishes his interests were unmistakably material. He wanted to assess the resources of the west coast of Africa and the possibility of a southern passage to India, and so he pressed his captains to go further and further. To do so, they had to overcome a couple of obstacles. The first was the fear of the unknown, especially given the frightening tales that were widespread and the reputation of the waters beyond Cape Bojador—its Arabic name, *abu khafar*, meant "the father of danger"—on the Atlantic coast of Morocco, which was said to be the point beyond which it was impossible to sail.

Alongside fearsome creatures and fanciful charts, the second obstacle that Henry's sailors faced on the open ocean was that their ships just weren't up to it. More than anything else, this fact made Portuguese seafarers wary of venturing too far south and facing northern head winds and adverse currents on the return voyage. Unlike Hanno's ships two thousand years earlier, the ones Henry sent forth had no oarsmen and depended on sail alone.

Since staying at home—the safest solution—was no option, the obvious way of dealing with the limitations of the technology was to improve it. Which the Portuguese did, adapting the lateen sail to the sterner requirements and longer distances of the Atlantic. The result, which revolutionized European seafaring, was the caravel, a ship that combined square and triangular sails on three masts. Columbus had them, and so did John Cabot. The standard caravel was relatively small, often only fifty or sixty feet long, with square sails on a couple of forward masts and a lateen sail at the rear. And it was well adapted to sailing out onto the seven seas and, more importantly, also returning from them. In 1434, Gil Eanes, sailing for Henry the Navigator, ventured past the dreaded Cape Bojador; and on his return, he not only disposed of the myths of boiling water and fierce monsters but demonstrated that a caravel could get back home relatively easily by taking a turn on the open sea, heading northwest from the Canaries to the Azores and back east from there to Lisbon.

Once Eanes had shown the way, progress was swift. By the time of Henry's death in 1460, ships he sent out had passed the Cape Verde Peninsula (in what is now Senegal) and were within 10 degrees of the equator. Then, in the winter of 1487–88, Bartolomeu Dias rounded the Cape of Good Hope, which he called the Cape of Storms; but his king—John II of Portugal, safe at home—changed the name to encourage the next wave of explorers. The entire west coast of Africa was finally added to the European map of the world, and a new sea route to India and the East lay open. Vasco da Gama soon followed, reaching the Malabar Coast of India in 1498, and his achievement was celebrated in one of the great modern epics, *The Lusiads* (1572), by the Portuguese Luís de Camões. He has his sailors land on the Island of Love in the middle of the Indian Ocean on their return. Like Odysseus, and indeed like sailors from time immemorial, they needed rest and recreation. Maritime enterprises, and stories and songs about sailing the seas and sometimes stopping on islands, have gone hand in hand from the beginning of seafaring time.

Chapter Three

The Origin of Islands

Ocean Bottoms and Volcano Tops

"Iceland is a large island situated on the verge of the Arctic Ocean [. . .] The country in general is mountainous; but in some districts, particularly those extending from the south to the north coasts, nearly through the centre of the island, there are extensive plains, for the most part dreary wildernesses, and covered with herbage only near the sea, or where morasses have formed. The highest mountains are covered with perpetual snow, and are chiefly, if not all, volcanic [. . .]

The greatest curiosities which Iceland presents are the springs of hot water [. . .] and the magnificent and tremendous explosions of the geysers [. . .] On the sulphur mountains, in the district of Guldbringè, are a number of jets of steam, and natural cauldrons of black boiling mud; and there is scarcely a district in the whole island without such indications of subterraneous heat, which must occasion the most singular contrast with the winter snows and ice, through which, at that season, they rise [. . .] The volcanic formation of Iceland is perhaps the most extensive in the world, covering an extent of at least 60,000 square miles."

—*Edinburgh Encyclopaedia* (1830)

"I NOW THINK that almost the whole world has learnt about the nature of the mountains in Iceland [. . .] On their peaks the snow is almost everlasting, yet in their depths sulphurous fire blazes away incessantly without consuming itself. Those who approach too close are easily suffocated by the quantity of dust and embers gushing out, and most of all when glowing-hot chasms appear in many places filled with the ash of burnt-up mountains and valleys. Here, growing silently from below, the sulphur increases as though following a natural cycle, until these ravines become ready for burning again [. . .]

Iceland is an island lying beneath the celestial Arctic Pole; it is mainly exposed to the wind Circius and close to the Sea of Ice. For this reason it deserves the name Ice Land (*terra glacialis*) or remotest Thule, which none of the ancients has failed to mention [. . .] Praise is due to this island for its unusual marvels. It contains a rock or promontory which, like Etna, seethes with perpetual fires. It is believed that a place of punishment and expiation exists there for unclean souls. Undoubtedly the spirits or ghosts of the drowned, or of those who have met some other violent death, are to be seen there."

—Olaus Magnus, *Description of the Northern Peoples* (1555)

In many ways, Iceland—spelled *Ísland* in Icelandic—is the archetypal oceanic island, with volcanic geology, strange geography, curious flora and fauna, and a unique culture. The sea around it ensures both its isolation and its community—and its island credentials as a place for a bird to land. True to form, millions of them arrive on the island every year, foremost among them the arctic terns whose return from their annual fifty-thousand-mile migration from the Arctic to the Antarctic and back is celebrated by Icelanders in the spring. The tern's Icelandic name, *kria*, catches the sound of its cry (or maybe, from the tern's point of view, its song), though in another island signature it is also a statement of defiance by this fiercely territorial bird. Other birds, such as auks and puffins and kittiwakes and fulmars and gannets, are attracted to the sea cliffs where they breed and lay eggs (a gift of nourishment to Icelanders and various animals), and some birds come to the island from their arctic habitats for the relatively mild winters on its southern lowlands. Ptarmigan waddle in the heather, prey for falcon and fox, while a variety of gulls fly around the shore, and ducks and geese swim inland. Oystercatchers and sandpipers wade on the coast, and ravens are everywhere, scavenging dead seals and whales and generally making mischief, the way ravens do.

Turtles, other reptiles, marine mammals, and of course fish swim all around Iceland. The only land mammal that has been on Iceland before humans is the arctic fox, which must have trotted across the frozen water during the last Ice Age, around ten thousand years ago. It stayed, feasting on berries and seaweeds, birds and their eggs, and small sea creatures. When humans eventually came and established farming communities, the foxes feasted on domestic animals, and so, for a long time, there was a bounty on them. Nowadays, they are hunted for their fur.

Over several millennia, other mammals arrived, including rabbits and reindeer, with the occasional polar bear also stepping off an ice floe. These white bears were tamed during the Middle Ages (as brown bears were in the rest of Europe), but in the past centuries they have been rare visitors—only about fifty were recorded in the last hundred years. Indeed, land animals of any size are relatively scarce on Iceland, with so much of the island under glaciers or scoured with lava, and some of it a northern desert. But there are ancient breeds of sheep and cows and horses there, found nowhere else and protected by prohibitions against animal import that have been in place for a millennium. Few trees survive, though when the Vikings arrived in the ninth century CE, birch trees covered a third of the island; now juniper bushes are much more common. Fish is the staple of an Icelander's diet, and food for many marine mammals; and the ocean is full of cod, which is the most important commercial fish, though herring comes close. Halibut, plaice, and lemon sole are among the most popular flatfish. Shark was also in demand during the nineteenth century, its oil lighting up Europe.

Iceland is a place of surpassing, surprising, and often—around the volcanoes and geysers and glaciers—terrible beauty. The earliest people to come there were probably wandering Irish monks, on their way to heaven. For early Christians, islands in the North Atlantic had an appeal similar to deserts in the Middle East—places of solitary trial and tough love, with the sea around the northern isles providing a wilderness of wonders, a welcome to wandering spirits, and a reminder of human limitations. The Viking seafarers arrived from Norway, led by two adventurers—identified in the medieval Icelandic *Landnámabók*, the *Book of Settlements*, as Ingolf and Hjorleif—along with a congregation of their servants and slaves. By that time there seems to have been little evidence of the original Irish settlers, and the Vikings soon established themselves. They might not always have been the nicest people to have around, but Icelanders have turned the historical account of rough-and-tumble Vikings into an heroic heritage; and it passes for the truth—at least on the island. The Icelandic novelist and Nobel laureate Halldór Laxness, on the occasion of the eleven hundredth anniversary of the continuous chronicled settlement of Iceland in the year 874, suggested that Icelanders have descended as much from books as they have from men. He was referring to the great sagas that have a central place in their storytelling tradition, and which have become true within one of those insular narratives of nationhood that we all live by. Laxness poked fun at the invention of Icelandic genealogies that traced pedigrees back to Homer's Troy, and he described the storytelling of the Icelandic scholar Arngrímur Jónsson, who, in the late sixteenth century, set out to counteract the belief that Icelanders were the descendants of Viking robbers, murderers, and slave owners by creating another version of history—straight from stories and songs of medieval Iceland like the great *Njal's Saga*, in which Icelanders come from a long line of aristocratic heroes, noble commoners, and poets. In due course, Arngrímur's history changed northerners' sense of themselves, and inspired nationalist Scandinavian uprisings. It also encouraged assumptions of Nordic superiority. So be it, said Laxness, sadly. The Swede Olaus Magnus, like Arngrímur writing in the sixteenth century, described how Icelanders "have a history of glorious feats. Even today they write down the deeds of their own age, which they recall in rhythmical songs, carving them on headlands or rocks, so that none may be lost to posterity."

Perhaps making up and marking down stories is an island habit; but if so, we are probably all islanders. The French Renaissance writer Michel de Montaigne (whose essay "Of Cannibals" inspired Shakespeare's island play *The Tempest*) punned on his own name by asking, "What truth is that which these Mountaines bound, and is a lie in the World beyond them?" Islands and mountains are the quintessential storytelling sites, set apart from the world and its realities—as are all stories, in some sense, insulating us in the many-roomed house of the imagination from the reality outside. The Majorcan storytellers' "It was, and it was not" is as intentionally contradictory as the "Once upon a time" with which we open story time for our children, since we mean "right now." Believe it, and not—because the story is both true, and not true. Especially on an island. Most especially, perhaps, on an island of fire and ice.

"This is an island and therefore / Unreal," wrote the poet W. H. Auden about Iceland. Its unreality has had a hold on imaginations for a long time, and the idea of Iceland as an "imagined" island goes back a couple of thousand years, associated in some early accounts with Ultima Thule, the land at the northern end of the earth. Sometimes Iceland was imagined in grim terms, as in a twelfth-century description (by the Danish writer Saxo Grammaticus) of how "at certain definite times an immense mass of ice dashes onto the rocky coast, the cliffs can be heard re-echoing, as though a din of voices were roaring in weird cacophony from the deep. Hence a belief that wicked souls condemned to a torture of intense cold are paying their penalty there." Visitors in the eighteenth and nineteenth centuries invented a more genial prospect, describing Iceland as pure and uncontaminated and romanticizing its social and economic wonders —in much the same way as they did with the islands of the South Pacific—with some (including Theodore Roosevelt) celebrating it as the home of democracy, while others (William Morris, for instance) were proposing it as a universal utopian model.

Which is to say that utopias don't always look the part, especially if they are islands in the North Atlantic. Not too far from Iceland, the island cluster of St. Kilda in the Outer Hebrides (west of Scotland) had been for at least two thousand years a home to humans, who lived there in a peaceable (some would say paradisal) manner that was celebrated by Lachlan Maclean in 1838 (and is now just a memory, since the last few inhabitants left St. Kilda in 1930): "If St. Kilda is not the Utopia so long sought, where will it be found?" Maclean asked. "Where is the land which has neither arms, money, care, physic, politics, nor taxes? That land is St. Kilda. No taxgatherer's bill threatens on a church door—the game-laws reach not the gannets. Safe in its own whirlwinds, and cradled in its own tempests, it heeds not the storms which shake the foundations of Europe [. . .] and cares not who sways the British sceptre. Well may the pampered native of happy Hirt [the Gaelic name of the island] refuse to change his situation—his slumbers are late—his labours are light—his occupation his amusement. Government he has not—law he feels not—physic he wants not—politics he heeds not—money he sees not—of war he hears not. His state is his city, his city is his social circle—he has the liberty of his thoughts, his actions, and his kingdom and all the world are his equals. His climate is mild, and his island green, and the stranger who might corrupt him shuns its shores. If happiness is not a dweller in St. Kilda, where shall it be sought?"

Beyond the fire and ice and the old sagas and even older parliament (the Althing, established by the Vikings) for which it is famous, Iceland represents a geological contradiction, simultaneously creative and destructive. In contrast to Jamaica, though like Tahiti, Iceland is a product of the separation rather than the sliding by or coming together of the tectonic plates that make up the earth's brittle outer layer. Its situation is complicated, however, for it is located astride an oceanic ridge that runs right up the middle of the Atlantic, all the way from Antarctica to the Arctic. Being in this geological borderland makes Iceland either the most easterly island in North America, or the most westerly in Europe; it is certainly an island in between—another island archetype. The sides of this mid-Atlantic submarine ridge are still spreading by about an inch a year as the seafloor moves apart. But enormous amounts of the ever-fresh volcanic lava rising from the earth's interior (which also created Iceland in the first place) keep the widening rift filled and the island intact. Somewhere in the world, almost every day, islands expose volcanic forces, breaking down and building up at the same time; and everyone on a volcanic island like Iceland lives constantly with this contradiction.

Scientists tell us that the earth, our island in space, was battered about for billions of years by meteors before it built up mass and generated heat that melted some of the metals, which were pulled to the center by gravity and solidified into an inner core surrounded by liquid. A thick mantle of molten rock developed around this core, and the mantle now makes up the large majority of the earth's volume and mass. A crust formed on the surface of our planet, producing gases that eventually precipitated into water, covering all of the crust until continents began to form.

The continental crust is relatively thick and consists mostly of granite, while the oceanic crust is thinner and composed of dense basalt. The crust and the top part of the mantle together make up the so-called lithosphere (from a Greek word meaning "rocky") in the form of enormous rigid plates that vary in size, consistency, and thickness. These plates (also called tectonic, or "building," plates) are either oceanic or continental (such as the Pacific Plate and the North American Plate).

Altogether there are ten or so large plates and a number of smaller ones, which all ride on top of a much warmer rock layer called the asthenosphere (from the Greek word for "weak"). The asthenosphere, which is part of the mantle of the earth, is under such intense heat and pressure that it behaves like thick honey, and the tectonic plates above it are constantly shifting as they shuffle about and shove each other, causing earthquakes and volcanic activity at their edges.

Iceland is not only located astride the Mid-Atlantic Ridge, it also sits on top of what geologists call a hot spot, where magma rises up from the earth's deep interior and melts through the crust. All ocean islands live on borrowed time, most sinking below the surface within five or ten million years, or drifting as much as six hundred or seven hundred miles during that time because of the movement of the tectonic plates. If there weren't such a hot spot under Iceland, the island would either split and the separate halves would drift in opposite directions, or it would cool and sink, since it is made up of heavy ocean rocks of volcanic origin.

Iceland is the most famous volcanic island in the Atlantic, because of both the size and the frequency of eruptions. The archipelago of the Hawaiian Islands is the best-known hot spot in the Pacific, and is home to continual and voluminous eruptions, with sheets of highly fluid lava common to this day. Its volcanic history is evident in its ancient landscape. The island called Hawaii, also known as the Big Island, is home to several volcanoes, one of which is the genuinely highest mountain in the world: the seasonally snow-capped Mauna Kea (White Mountain), measuring over thirty-three thousand feet from top to (sea) bottom, compared to Mount Everest, whose height is twenty-nine thousand feet. Mauna Loa, also on the Big Island, stands next in height to Mauna Kea; both of them broke the surface of the sea about fifty million years ago. Of course, their height is calculated from the ocean floor, and we don't usually measure mountains on the mainland that way; but even when we start from "sea level," we recall the relationship between mountains and islands and the sea.

Everywhere on the planet, and especially in the geologically volatile Pacific, ancient story and song cycles are filled with accounts of islands rising out of the sea. In Tonga, there are stories of the "times of darkness" that almost certainly refer to volcanic eruptions, and one remarkable image of the wife of the legendary Hawaiian navigator Makali'i—his name means "Eyes of the Chief"—sounds like it could also be a description of a volcanic eruption. "Her skin was as red as fire, on coming out of the house, her beauty would overshadow the rays of the sun, so that darkness would cover the land, the red rain would be seen approaching; the fog also, and after these things, then the fine rain, then the red water would flow and the lightning play in the heavens. After this, the form of Malanai-kuaheahea would be seen coming along over the tips of the fingers of her servants, in all her beauty. The sun shone at her back and the rainbow was as though it were her footstool."

Of course, islands also disappear, sometimes sinking under their own weight or submerging when the sea level rises, as it does from time to time—and appearing again when the waters recede. Atlantis, real or imagined, figures prominently in this storyline. Great Britain has been up and down several times in its geological history. And like the Hawaiian Islands, the Channel Islands off the coast of California were more numerous during the last Ice Age, several of them part of a single island larger than the current archipelago combined. And we may be about to witness another cycle of submerging islands in the current round of climate change. This is the story of the world, and of its islands. Climate change is one part of it, and has been going on since the beginning of time, long before humans arrived to accelerate the process. Volcanoes are another.

Iceland's record of volcanic activity began millions of years ago, but a reminder of its perennial presence occurred in 2010, when the Eyjafjallajökull volcano erupted, spreading ash over Europe and North America and disrupting air travel around the world for a fortnight. On the island there are memories of other eruptions. In 1783, a volcano caused the opening of a fissure over fifteen miles long, releasing lava that covered two hundred square miles, completely filling two valleys and creating cones of volcanic cinder similar to those that have been identified along mid-ocean ridges. Even deadlier than the lava, however, were the gases released back then, which created a blue haze over much of Europe and choked most of the livestock on Iceland, producing massive famine. Though particularly devastating, this was only one of many, many volcanoes that have shaped, or mis-shaped, Iceland, and at times exhausted its residents. In 1875, a volcanic eruption prompted a mass migration to the United States and Canada, with the result that the largest concentration of Icelanders off the island now live in the town of Gimli, Manitoba.

Such island catastrophes do not always lead to a permanent exodus, however. Tristan da Cunha in the South Atlantic, over seventeen hundred miles from the continent of Africa, is one of the most remote island groups in the world. In 1961, an active volcano forced the evacuation of the eighty or so families who lived there, and they were relocated to southern England. But when an expeditionary task force went there a year later to inspect the situation and reported that the main island settlement was mostly unaffected—it's called Edinburgh of the Seven Seas, after a visit by the Duke of Edinburgh back in 1867—almost all the residents returned.

Some say land on our planet first rose up above the water about a billion years ago in the form of a continent called Rodinia (from a Russian word meaning "motherland"). Others insist that there was a supercontinent called Vaalbara a couple of billion years before that, when life as we know it possibly began (though another story has it starting with chemosynthesis at underwater vents rather than with photosynthesis on land). Either way, sooner or later there was land, surrounded by water. The first island on island Earth.

But nothing on Earth is forever. Life is followed by death; water rises and falls; land comes and goes. On that, everyone agrees; and in the course of time Rodinia broke into pieces. In its end lay the seeds of a new beginning, however, for the pieces converged into other continental formations, which in turn came together about 275 million years ago to create one island continent called Pangaea, a word that means "the entire earth." Pangaea, too, was surrounded by water, an ocean named Panthalassia.

Then, about 200 million years ago, the landmass split apart once more, and the water rose, creating two continental formations, Gondwana in the south and Laurasia in the north. In due course, they sponsored the continents we recognize today, though there was much drifting and shifting of plates in the meantime. The superocean, in turn, was divided into several smaller ones, offspring of the original and still covering much of our blue planet.

Today, most of the earth's land above water is comprised of the continents, but islands also count for a lot of it. How many islands are there in the world? Somewhere between two hundred thousand and—if rock outcrops and tidal islets are included—two hundred million. It all depends on how we define an island. The difference between islands and continents, too, is a matter of definition—a mix of geology, geography, and politics. Size matters, of course; but continents are mainly defined by their origin. They were once part of those early supercontinents, and include evidence of both the makeup and the breakup of the supercontinental masses. Because of their heterogeneous geological history, continental rocks (which are mostly granite) tend to be lighter than those on the ocean floor, and therefore split and shift more easily. This gives continents a geology that is typically complicated, in contrast to the simpler geology and heavier rocks—usually volcanic basalt along with coral limestone—of most ocean islands. Of course there are anomalies. Japan and New Zealand and the Seychelles are geologically similar to continents, except in size (with granite rock formations in the Seychelles nearly a billion years old). Other islands, like Ireland and Newfoundland, display the effects of volcanic action in the shallow waters of the continental shelves, with further build-up by sediment brought down by coastal rivers, smoothed by glacial scouring. But a lot happened before that, with these two North Atlantic islands once part of the same landmass, until the closing and reopening of the Atlantic hundreds of millions of years ago reshaped Newfoundland's geology, with its hills connected to America's Appalachian Mountains.

But most of the world's ocean islands share a common heritage, having been created when submarine rocks were raised up either by the action of underwater volcanoes or by the (less dramatic) drifting of the tectonic plates—and often something of both. Jamaica, for example, was formed when the Caribbean Plate moved east, the North American Plate west, and they rubbed up against each other. The overall result was the Greater Antilles. And although Jamaica has had some violent geological moments—such as in 1692, when an earthquake devastated Port Royal—its physical history is relatively sedate, like that of neighboring Hispaniola and Puerto Rico and Cuba, each of which shows evidence of the slow slipping and sliding of the plates and the gradual accretion and erosion of the rocks on the earth's surface. In contrast, islands in the Lesser Antilles have a much more dramatic record—like the rhythms of calypso and soca compared to ska and reggae. On Martinique, a volcanic explosion in 1902 killed thirty thousand people; and as recently as the 1990s, a volcano on the island of Montserrat, dormant since the sixteenth century, destroyed the capital and forced half the population to evacuate. That volcano is still active, and there are many others, both on the islands of the Lesser Antilles and on the ocean floor in that part of the Caribbean, which either erupt regularly or are a constant threat. One of the latter, called Kick 'em Jenny, between the islands of Grenada and Carriacou, has erupted at least a dozen times in the last century. It is now just five hundred feet below the surface, and sooner or later a new island will probably emerge, joining the others in the arc of ocean volcanoes that make up the islands of the eastern Caribbean.

New islands are relatively common, especially in the volcanic ocean regions of the world; but as with so much on Earth, they are often here today and gone tomorrow. Near Sicily, an island called Ferdinandea is now twenty feet under water and a danger to shipping. It was originally identified above sea level in 10 BCE but then disappeared beneath the surface, only to emerge again in the 1830s, when it was claimed by the British and named Graham Island. But neither the name nor the island stayed for long, and Ferdinandea is for the time being just a silhouette beneath the surface. Kavachi, in the Solomon Islands, has appeared at least nine times in the last century, but, in a peekaboo fashion, never stayed above the surface for more than a few days. Falcon Island, in the archipelago of Tonga, has come and gone a dozen times in the past couple of centuries, occasionally reaching over four hundred and fifty feet in height, with a length of three miles and a width of half that. A description of Falcon Island by an observer in the 1880s catches the fearsome beauty of such uprisings. "The scene was most magnificent, great volumes of steam, of carbonic and sulphurous gas [. . .] being shot forth from many jets out of the sea, in a direct line of over two miles [. . .] to the height of 1,000 feet and more, then expanding themselves in all directions, in clouds of dazzling whiteness, and assuming the most fantastic shapes; sometimes presenting themselves as a mountain of wool, the tips of which were fringed with gold, caused by the rays of the setting sun, then again occasionally forming into a white cauliflower head of snow whiteness, backed by clouds of intense darkness formed of dust and ashes mixed with watery vapour, which the wind was carrying down for miles on the distant horizon. As the heavier matter kept continually falling, it gradually raised in height the new-made island."

The Pacific is certainly not the peaceful place it seemed to Magellan when he first sailed into its waters and wanted to give it a name, for in geological terms its floor is dramatically unstable, with constant volcanic activity and continual slipping and sliding of tectonic plates. In some ways it is the most dangerously unstable part of our planet. Consequently, a large percentage of the world's islands are found there, surrounded by the so-called Pacific Ring of Fire, a horseshoe-shaped zone of convulsive volcanic activity where heavier oceanic plates slip below their lighter continental counterparts, resulting in tension, wrenching, and release in the form of deep earthquakes and explosive surface activity—

one example is Bougainville Island (in the Solomon archipelago), which has three active volcanoes and as many as forty earthquakes a year. The threat of tsunamis from underwater earthquakes is constant, as the people of Japan and the islands of Indonesia know all too well.

The forces required to raise islands from the ocean floor have often defied description, and the association of volcanic and earthquake activity with supernatural powers has been a constant in human history. For islands don't only highlight the natural geology of the world, they also illuminate its supernatural geography, with flood and fire regularly associated with the powers of the gods, both beneficent and malign. The convulsions that characterize island formation—the volcanoes and earthquakes that intermittently create and destroy so many of them, as well as the slow but sure effects of winds and waves—have found human expression in a pantheon of gods and spirits who were routinely imagined as residing on islands. For millennia, shamans and wizards and priests and other custodians of supernatural power took up residence there, either to be near the source of what they identified as divine power or to mirror its conditions.

For a long, long time, people around the world have offered explanations in their creation stories for the awesome forces of nature that build up or break down the world—the geological forces—and also cause the emergence of life-forms—the forces we have come to call biological, or evolutionary. Many of these storytellers come surprisingly (or perhaps *not* surprisingly) close to the explanations of modern geological and natural science. One reason may be that, like the scientists, they often center their tales around islands. The one thing that everyone agrees, across cultures and throughout the ages, is that we are dealing with forces almost beyond human understanding, and far beyond human control. Islands—their sometimes spectacular formation, their often strange shape, their occasionally sudden disappearance—prompt a unique kind of wonder that appeals to religious and scientific as well as philosophic and political imaginations. And so what Aristotle called the "final cause" of islands, their design or purpose in the grand scheme of things, is usually identified somewhere between the material forces of geology and the spiritual forces of creation—a nice counterpart to the forces that shape our own lives. Indeed, geology itself has often sponsored connections between the natural and the supernatural, and between science and religion.

When Charles Darwin encountered the devastating earthquake near Concepción in Chile in 1835, he noted in his journal that "we can scarcely avoid the conclusion that a vast lake of melted matter is spread out beneath a mere crust of solid land [. . .] Nothing, not even the wind, is so unstable as the crust of the earth." Indeed, Darwin began his career as a geologist, at a time when the origin of islands was mostly mysterious. Throughout his life he followed their geological and geographic exploration with careful attention; and as his interests turned to natural history, he recognized that the origin of islands and the origin of species were intimately related, and that both could be explained by looking at things that were still happening. Speculating that islands must have emerged by the agency not only of falling sea levels but also by the rising of the ocean floor, he came to the conclusion that it was volcanic activity underwater that created most ocean islands, which then very slowly would disappear again beneath the surface of the sea, due to a combination of their own weight and the actions of wind and water, with some of them being transformed by coral activity into barrier reefs and atolls.

Although Darwin is often seen as the architect of the antagonism between science and religion, in fact he made them fellow travelers, each contemplating the almost unimaginable forces of island formation in their own way. And he had his own fellow traveler on his voyage to the Galápagos in the person of his friend Robert FitzRoy, who was both highly inquisitive and deeply religious, and on both counts well acquainted with the wonders of the world. FitzRoy was the captain of HMS *Beagle* on the five-year voyage of discovery from 1831 to 1836 that ultimately led to Darwin's *On the Origin of Species by Means of Natural Selection* (1859). Appropriately, FitzRoy's legacy lives on, not only for his leadership and companionship but also for developing the first accurate storm warnings for ships—he had observed that a sudden drop in pressure, signaled by the barometer falling rapidly, usually preceded bad weather. In later years, he founded the British Meteorological Office and designed easy-to-use barometers for fishermen, providing an instruction manual in rhyming couplets to make it easy to remember. During the voyage of the *Beagle*, FitzRoy also tried out a new system for classifying the force of the wind that had been devised by his mentor, the British admiral Francis Beaufort; it was the basis of the modern Beaufort scale. Lastly, his name is remembered in the regular BBC Radio weather broadcasts where listeners who might never take to the sea hear the daily litany of good weather or gale warnings in the North Atlantic sea areas of "Fair Isle, Faroes, Iceland, Hebrides, Rockall, Shannon, Sole, FitzRoy."

When Darwin returned from his journey, he reflected on the "pleasure of the imagination" that islands inspired in him—and he had visited plenty of them. He admitted that even after all his geological speculations, he still had an "ill defined notion of land covered with ocean, former animals, slow forces cracking surface." But it was all "truly poetical," he concluded. It was the poetry as well as the prose of islands that inspired Darwin and other scientists of his time—and still inspires us, for a combination of imaginative wonder and intellectual wondering has generated much of our current understanding of both the physical (geological) and the natural (biological) history of the world. This understanding did not come easily, for it was only with the renewal of old tensions between religious and scientific explanations in the late nineteenth century that a new appreciation of the relationship between nurture, nature, and the supernatural emerged.

The disagreements showed themselves, among other times and places, in Jamaica three centuries ago, where the slip-sliding scene created by the earthquake in 1692 (and still visible today) in the coastal village of Port Royal not only reshaped the land but also renewed scientific and religious bewilderment over the forces behind such cataclysms. The earthquake turned the water-saturated sand on which the village was built into a kind of liquid, and created a submarine landslide that left only a few buildings standing, at a crazy angle. But Port Royal had been known as "the wickedest city in the world" because it harbored pirates and privateers, and many thought the Port Royal earthquake was divine retribution. "There never happens an earthquake," said the Boston Puritan Cotton Mather about the Port Royal cataclysm, "but God speaks to men on Earth."

This was not an unusual response. After two earthquakes devastated Lisbon in the 1750s, people began to think they were being punished for some terrible wrong they had done. One such cataclysm might be geological, some speculated; two or more were definitely theological. The earthquake that had occurred in Concepción in 1835 prompted wide-ranging speculation that certainly influenced Darwin's own thinking about the origin of islands—and of species. Some thought the earthquake had been caused by an Amerindian witch, who had been so insulted by the people that she went up into the mountains and plugged the volcanoes. To Robert FitzRoy, it was an act of God. But wonder comes in various guises, and does not always fall neatly into scientific or religious categories. It was FitzRoy, after all, who had given Darwin a copy of Charles Lyell's *Principles of Geology* (published in three volumes, 1830–33), which Darwin took along on his journey; and Lyell was the prophet of modern geology, to whom Darwin dedicated his *Journal* of his voyage on the *Beagle*. Sooner or later "every valley shall be exalted, and every mountain and hill shall be made low, and the crooked shall be made straight," said Isaiah. So did Charles Lyell.

Religious or scientific, mythical or anecdotal, almost all the accounts of the creation of the world carry in them the seeds of the world's destruction, with the end often mirroring the beginning. They agree about the dynamics of generation and degeneration, though religion and science quibble over the order of service. Tales of fires and floods are familiar fodder for storytellers in both camps, and so are stories of the rising and falling of islands. They provide our most enduring image of the beginning of the world, when light was separated from darkness and land from water, and when human life was made possible. And of its ending, too. Some of us see the present melting of the polar ice pack—with the consequent loss of islands, from tropical atolls like the Maldives and Tuvalu to the ice islands in the Arctic—as a contemporary version of the biblical flood, brought upon us because we have done wrong. (It has, incidentally, also created considerable challenges regarding the geopolitical status—and the undersea mineral rights—of island nations that might sink below the sea.) This convergence of stories around islands—their fate and our fate inextricably interwoven—has been evident for a very long time, and reflects an awareness of the mysterious powers that are conjured up in both scientific and religious explanations of the beginning and end of the world, and its conduct in between.

There are regular attempts to reconcile these explanations. The British biologist J. B. S. Haldane who, in the 1920s, proposed a theory of the origins of life in what he called the "prebiotic soup" of the ocean, made a distinction that was friendly to the stories of both. Religion, he suggested, "is a way of life and an attitude to the universe." It brings us "into closer touch with the inner nature of reality. Statements of fact made in [religion's] name are untrue in detail, but often contain truth at their core. Science is also a way of life and an attitude to the universe. It is concerned with everything but the [inner] nature of reality. Statements of fact made in its name are generally right in detail, but can only reveal the form, and not the real nature of existence." Sometimes the different storylines reflect another kind of difference, between looking at the origin of things and looking for their purpose, between formal explanations of cause and effect and functional explanations of overall pattern and intelligible design. But as the natural scientist D'Arcy Thompson remarked in his great study *On Growth and Form* (1917), form and function are woven together in our understanding of the world like warp and woof. Which may be why we are so fascinated with the beginning and ending of islands.

When scientists in the nineteenth century got around to thinking seriously about islands, their rising and falling seemed to some the story of Earth itself, beginning with the original separation of land and water when the water fell and the land rose, and then the water rose and the land fell back—again and again. Like breathing in and out, as mythmakers had said for millennia, for islands offered an image of an earth that was alive. Indeed, "physiology" was the word that Darwin used when he began his work, bringing together organic and inorganic science; and his journal of the famous voyage of the *Beagle*, on which he had begun to formulate his theory of evolution, was titled *Journal of Researches into the Natural History and Geology of the Countries Visited During the Voyage of H.M.S.* Beagle *Round the World.* It was published in 1839 (twenty years before *On the Origin of Species*) and was followed in 1842 by *The Structure and Distribution of Coral Reefs.* Two further volumes of *Geological Observations* on volcanic islands and South America appeared soon after.

As he was writing up the account of his voyage on the *Beagle* for publication, Darwin concluded with a long meditation on island formation. Why are coral atolls common in the Pacific, but not in the islands of the tropical Atlantic, he asked himself. Because the Caribbean islands are still on the rise, he speculated, while the Pacific islands have, after a long period of rising up above the surface, been sinking at a rate that, while measurable, is on geological rather than Greenwich time. Scientists call it deep time, and it is less involved in sequences than in patterns. For it turns out that in the history of the earth, direction is not as important as we sometimes think. This has long been known by mythmakers but had also been observed by the eighteenth-century geologist James Hutton and described in his landmark *Theory of the Earth* (1795), in which he brought together the notions of time as an arrow and time as a cycle (terms made popular by Stephen Jay Gould) in a description of the geological design of the earth and its islands.

Darwin was trying to complete our understanding of that design, and his great achievement was to see evidence of deep time in natural as well as geological history. He began, as the earth and all its islands did, with geology. "We see in each barrier reef," he wrote, "a proof that the land has there subsided, and in each atoll a monument over an island now lost. We may thus, like unto a geologist who had lived his ten thousand years and kept a record of the passing changes, gain some insight into the great system by which the surface of this globe has been broken up, and land and water interchanged." A hundred and fifty years later, the influential marine geologist Henry Menard put it differently, noting that the precision of our measurement of such effects masks the mystery of their cause. "Islands are dip-sticks that record the changing distances from the sea floor to the ocean surface," he wrote. "Unfortunately, the record locally is the same if the sea floor goes down or the sea surface goes up, and the history of sea level changes is not necessarily informative about causes."

Darwin recognized—even though he couldn't fully explain it—the interdependence of physical (or geological) and natural (or environmental) causes; and he insisted that they were as obvious in the present as they were evident in the past, which brought his experience as a geologist into harmony with his instincts as a naturalist. The idea behind it was known as "uniformitarianism," and it had been argued by scientists such as Lyell and Hutton in opposition to what was called "catastrophism," which presumed that the physical and natural world was formed in a series of powerful independent events. Biblical accounts, read literally, tended to fall into the latter—the catastrophic—camp, while Darwin and most of his scientific colleagues believed in a uniform, universal, and continual process. But Darwin was almost alone in seeing the connection between natural and physical explanations.

The Inuit have a word for a mountain peak that rises above the continental ice. They call it a *nunataq*; and you can see such peaks throughout the Arctic. With a *nunataq*, as with an iceberg, it's what you cannot see—what is below the surface—that is both fascinating and fearsome. Sailors have always understood that islands are like this, mountains rising from the ocean floor—but until the past few years nobody had ever been down there.

Sounding the deeps must have been part of seafaring ever since the first ship ran aground; and from wall carvings we know that the Egyptians were using poles to establish the depth of the sea three thousand five hundred years ago. Herodotus writes about the Greeks using weighted lines a millennium later; and the use of "lead and line" is mentioned in Homer's *Odyssey* as a way of measuring both the depth of the sea and the nature of the seabed (from samples brought up by a lump of wax set into the base of the lead). The standard unit of measurement was the fathom, the reach of the line handler's outstretched arms—whence the sense that we can fathom (or understand) what is within our reach. Around the time of the birth of Christ, the geographer Strabo recorded that a sounding was taken to the depth of a mile; and in the story of the shipwreck of the apostle Paul on the island of Malta, there is mention of sailors taking regular soundings—though apparently not regular enough. In homage to Paul's sad story, the tune to the great naval hymn "Eternal Father, strong to save, / Whose arm doth bind the restless wave [. . .] O hear us when we cry to thee / For those in peril on the sea" is named "Melita," after the ancient name for Malta.

Until the nineteenth century, the technology didn't change much. There was a line and a sinker, sometimes with a hook (or bucket) to bring up rock samples, with a great deal depending on how much line you had on board. When Magellan attempted to sound the depth around the first islands he saw after sailing into the Pacific through the straits at the tip of South America, he didn't have enough line to hit bottom. Later innovations included copper wire instead of rope, but it tended to kink and snarl, so piano wire was introduced by the inventive Lord Kelvin (William Thomson) in the 1800s. All this echoes the story of the muskrat bringing up earth from the bottom of the sea to put on the back of Turtle Island, or the tale of the Polynesian hero Maui fishing up islands from the ocean.

And for a long while, fancy had a field day. When Jules Verne wrote *Twenty Thousand Leagues Under the Sea* in 1870 (the twenty thousand leagues—or fifty thousand miles—refers not to the depth of the ocean but to the distance traveled, twice around the world), he caught the imagination of young and old from all walks of life. But the creative imagination often coincides with scientific inquiry, and a couple of years later a great round-the-world oceanographic expedition set out from Britain on the *Challenger*, a naval vessel converted to a seagoing laboratory under the command of George Nares and the scientific leadership of Wyville Thomson. The *Challenger*'s mission was to sound the depths of the world's oceans. It returned to using hemp for sounding, but with significant advances in winding techniques, resulting in the discovery of the sea's deepest region: the Marianas Trench in the Pacific, where the seafloor is almost seven miles from the surface. (Over a hundred years later, when the technology was finally available, what is probably the deepest spot within the Marianas Trench—called the Challenger Deep—was measured.)

Between the *Challenger*'s return to port with its findings in 1876 (after a voyage of almost eighty thousand miles, taking dredge and dragnet samples in over three hundred fifty locations) and the later part of the twentieth century, a dozen research expeditions extended the inventory of the ocean floor, the most systematic being by the German vessel *Meteor*, which cruised the South Atlantic between 1925 and 1927, and the deep-sea drilling operations conducted by the American *Glomar* (i.e., Global Marine) *Challenger* beginning in 1968. But although something was learned about the geological composition of the ocean floor and its difference from the continents, the knowledge remained scattered until a set of new technologies allowed for a more precise mapping of the seabed and a more complete understanding of the geological history of both the ocean basin and the islands that rose to the surface.

The first of these was sonar and radar. The terms are acronyms from the 1940s, during World War II, for "*so*und, *na*vigation, and *r*anging" and for "*ra*dio *d*etection *a*nd *r*anging." But even with sonar and radar and magnetic scanning techniques, the pinpoint accuracy we now take for granted in both geological measurement and geographic mapmaking only became possible in the 1970s, after satellite technology was developed. Before that, neither twentieth-century scientists nor sailors knew much more about either the bottom of the sea or the precise locations of most of the islands on its surface than their predecessors did several hundred years earlier.

Much of our understanding of the geology of islands and of the earth is therefore very recent (just as it is with starry islands in space). Some of it is still in dispute, for the current steady-state model—in which the earth's plates shuffle around in a manner that maintains the size of the earth as constant—is challenged by a theory that argues that the earth is expanding. And almost all of our knowledge is incomplete, as science always is, for it relies on generalizations—sometimes elevated to theories—which don't always fit with the particularities of individual conditions. That said, there is no substitute for watching and wondering. Stars and galaxies are still imploding and exploding, creating and destroying planetary systems like our own, and oceanic islands are still rising like mountains from the ocean floor, propelled by volcanic fire and by the slip-sliding of the earth's lithosphere, and then disappearing again, either slowly or suddenly.

The explanation for the comings and goings of islands only began to take shape in Charles Darwin's day, and with his help. One person who was paying particular attention to Darwin's ideas about ocean islands was the American geologist James Dwight Dana, who led the United States Exploring Expedition (1838–42) to the Pacific islands (on a ship commanded by the irascible Charles Wilkes, said by some to be the original of Captain Ahab in Herman Melville's *Moby Dick*). Dana had read about Darwin's geological speculations in a newspaper in Australia in 1839, where he had stopped on his way to chart the island chains in the Pacific. Volcanic island arcs especially interested both Darwin and Dana, and both came to believe that volcanoes played a major role in the formation of the world's islands. They knew that their eruptions were often spectacular. Krakatoa was to provide a devastating example in 1883; but even that was small stuff compared to the Tambora volcano on the Indonesian island of Sumbawa, which had erupted in 1815 and spewed ten times more ash than Krakatoa into the air.

Dana was determined to establish a chronology for the geological history of oceanic island arcs. He observed a pattern, with active volcanoes at one end of island chains and deeply eroded volcanic rocks disappearing beneath coral reefs at the other; and he proposed a timeline, which included submerged islands. His account, with its identification of stages in the appearance and disappearance of ocean islands, turned out to be as important for geologists as the sequencing of organic species was for the naturalists; and like Darwin, Dana later combined his geological interests with natural history. When he sent Darwin a copy of his findings in 1849, Darwin embraced them immediately and responded, "last night I ascended the peaks of Tahiti with you." Darwin had first been there in 1835, and Dana in 1839, the only Western scientists to visit Tahiti since 1769, when Joseph Banks had accompanied James Cook on his first voyage.

However, Dana's mapping of island arcs left unanswered the deeper question of what actually coordinates volcanic and earthquake activity, though some geologists had proposed explanations, describing how breaks in the crust of the earth were caused by molten rock down below, rather than divine malice or mischief up above. And Dana's and Darwin's contemporaries were coming up with new insights. In 1858, the year before *On the Origin of Species* was published, the geologist Antonio Snider-Pellegrini was tracing coal seams and noted that the continents, most obviously the west coast of Africa and the east coast of South America, fitted together like pieces of a jigsaw puzzle. This was a startling (though not entirely original) observation, but nobody could explain how it all might have happened. Then the German meteorologist Alfred Wegener—who coined the term Pangaea to describe one of the supercontinents millions of years ago—took up Snider's idea and, in 1912, proposed that the continents had "drifted" apart. Unfortunately, his theory of continental drift was dismissed by most of the scientific community because nobody (including Wegener) could imagine the force necessary to make it happen. This skepticism continued for half a century, until geologists identified paleomagnetic patterns on the ocean floor that confirmed the movement of the oceanic as well as continental plates. Still, the *Standard Encyclopedia of the World's Oceans and Islands*, published in 1962, described the idea as "too controversial"; and in 1964 the *Time-Life Nature Library* stated categorically that "there are no known forces strong enough to move the continents around the earth, let alone split them into fragments [. . .] For these reasons the theory of continental drift has been abandoned by nearly all geologists."

Wonder, and wondering. The poetry of islands, like the poetry of the planets and the stars, has been with us from the beginning, providing inspiration for many discoveries in science, ancient and modern, and many myths, also both ancient and modern. Darwin had envisaged, in one of those "truly poetical" moods he wrote about, a vast array of flat-topped "mountains" rising up from the ocean floor—volcanic islands that were eroded to sea level, then eventually sank beneath the waves—stretching throughout the Pacific in patterns determined by volcanic activity; and Dana had proposed the same, independently of Darwin. It was a magnificent image; but it wasn't until a century later, when the vast terrain under the ocean was fully mapped, that Darwin's underwater poetry and Dana's elaborate calculations were confirmed and the origin (and extinction) of islands more fully understood.

It was the United States antisubmarine naval commander and geologist Harry Hess who (in 1945) identified the sunken islands that Dana and Darwin had imagined, in a continuous sonar survey of the ocean floor that he undertook during his wartime mission in the Pacific. He called these former volcanic islands, which were now underwater, guyots, after the flat-topped geology building at Princeton University that was named after Arnold Guyot, a nineteenth-century geographer. Hess also put forward the theory that the seafloor was broken by tension cracks that opened along fault lines and then spread to the side of volcanic ridges. The chronology that Dana had proposed was taken up when another geologist, Lawrence Chubb (who had retired to Jamaica), noticed that these underwater sea ridges slowly changed location, illustrating what he called "earth-movements." This wonderfully suggestive geological phrase reminds us how often the modern sciences of geography and geology have turned to organic metaphors to describe the history of the physical earth, and how these metaphors have coincided with the god-fearing myths of many ancient (and some modern) peoples. In Mozambique, for example, earthquakes are said to happen because the earth is seen as a living creature, with the same kind of problems people have. Sometimes it gets sick with a fever and chills, and begins to shake. We can dismiss this as fanciful, if we want; but the fact that Hess characterized his groundbreaking paper on "The History of Ocean Basins" as "geopoetry" is a measure of how mysterious much of this has been, even for scientists.

In the 1960s, the Canadian geophysicist J. Tuzo Wilson added another chapter to this story, one that transformed our understanding of oceans and continents and the forces that continue to shape the islands in between. He had been schooled by Harry Hess (his graduate teacher at Princeton) and by others who were investigating the geology of the ocean floor; but Wilson began with islands on the surface, observing the differences in volcanic activity among the Hawaiian islands on an airplane flight over the archipelago, and speculating about ancient rock formations while sailing in the summer among the islands of the Great Lakes (where some of the oldest rocks on Earth are found) in his beloved Chinese "junco," or junk—an island-hopping, ocean-crossing sailing boat whose design goes back over two thousand years. Wilson took the theories of continental drift and seafloor spreading and combined them into one theory of plate tectonics. The plates drift everywhere on Earth, and when they spread and slide by each other they create fractures on both the continental surface and on the seafloor. When they drift together, part of one plate (usually the oceanic plate, since the continental plate is lighter) disappears beneath the other in what geologists call subduction, in the process creating deep ocean trenches (like those in the Pacific) and high mountains like the Andes and the Cascades on the west coasts of South and North America, respectively.

Wilson predicted three types of boundaries between plates: the trenches, or subduction zones, where plates converge and one plate is drawn under another; mid-ocean ridges, also called spreading centers, where plates diverge; and fractures in the seafloor (or occasionally on land, such as the San Andreas Fault on the Pacific coast) called transform faults, where the plates slip by each other. Since tectonic plates are horizontally rigid, identifying the speed and direction of drift for a few points on a plate makes it possible to chart the movement of every point on that plate. Nothing is absolutely stable, and everything is in relative motion, albeit at the speed of a turtle. Wilson's insight was a counterpart to the new knowledge of genetics that transformed natural history, and it put the deep time of geology into conversation with the disruptive and productive mutations of evolutionary theory. Knowing that the drifting and shifting he described would be hard for his colleagues to imagine, Wilson came up with a paper model—like origami, the classic folding craft—to demonstrate how these enormous mobile plates are connected by ridges and trenches and fault lines. When he saw puzzled faces he would pull out his parlor trick, a modeling of a mystery of the sort that great scientists specialize in.

Inspired by Wilson, other scientists finally began to understand both the cause and the chronology of the changing features of continents and oceans, and the forces required to bring about the observable shifts of the earth's surface. And finally, too, they were back to the beginning of the story in myth and legend from time immemorial, that of the earth as alive, with pieces of it coming together and breaking apart in a sequence of seismic and structural transformations whose rhythms transcend the categories of the organic and the inorganic. The belief in a living Earth is much embraced today—but it is also an ancient faith, as the myth of Gaia and thousands of other stories around the world remind us. And islands continue to come and go, like life itself.

There is an island in the North Atlantic reminding us of this right now. It is called Surtsey, and it started out the way islands usually do—underwater. But on November 14, 1963, a plume of ash and steam and gas rose above the surface of the sea just off the southwest coast of Iceland, rising over two miles into the air. Over the next few months, a series of volcanic eruptions threw up pyroclastics, rocks broken into pieces by fire and water. Winter storms almost washed away the loose fragments on the low-lying island; but six months later, in April 1964, the accumulated volcanic rubble separated the center of the new island from the sea, and lava flow covered the loose rocks, protecting the island—for a while at least—from further erosion by wind and water. A new island was born.

It happens all the time, all over the world. But this birth was unusually public, with the people of Iceland watching (and knowing all too well the forces of volcanoes). The appearance of this new island coincided remarkably with the new theories about the earth's breakdown and makeup, theories that had been encouraged by other volcanic islands like Hawaii as the science of geology caught up with the imagination of storytellers. The island was baptized Surtsey after the mythical Icelandic giant Surtr, who figures in ancient tales as the god of fire from the underworld; and in its way Surtsey symbolizes the geological and natural as well as the imaginative history of islands.

A half century later, plants grow on Surtsey from seeds borne there by the wind and the waves, and in the droppings of birds that visit and sometimes stay. In the words of the American-Icelandic writer Bill Holm (whose surname means variously "hill," "sea," or "island" in the old languages of the north), "oceanic islands start from scratch—as indeed the planet itself did—with nothing, except the wind and bird shit which eventually carry a little life onto the cooling lava." Humans are scarce on Surtsey—none live there, at least not yet—because scientists want to watch its physical and natural evolution carefully; and humans, who for some reason are often considered "unnatural" island visitors, tend to confuse the picture. Volcanic forces continue to build below the sea's surface, while wind and water work to break up the island above. It is the oldest story in the history of the world.

Chapter Four

The Origin of Species

Island Plants and Animals

"Galapagos, or Tortoise, or Enchanted Islands, are the names which have been given to a cluster of desert islands situated in the Pacific Ocean [. . .] They were first discovered by the Spaniards, and have been since visited by [William] Dampier and [George] Vancouver. The southernmost island is about 4 miles in circumference, and the northernmost about 1 1/2 miles. Most of these islands [about fifteen substantial islands and a number of smaller ones] are flat, and tolerably high. Four or five of the most eastern are rocky, hilly, and barren, producing nothing but some shrubs on the shore. Others of this cluster produce trees of different sorts; and in some of the most western of the group, which are 9 or 10 leagues long, and 6 or 7 broad, large and tall trees, especially mammee trees, grow in extensive forests. In these large islands, rivers are of a tolerable size."

—*Edinburgh Encyclopaedia* (1830)

"NOTHING COULD BE less inviting than the first appearance [of Chatham Island in the Galápagos]. A broken field of black basaltic lava, thrown into the most rugged waves, and crossed by great fissures, is everywhere covered by stunted, sun-burnt brushwood, which shows little signs of life. The dry and parched surface, being heated by the noonday sun, gave to the air a close and sultry feeling, like that from a stove: we fancied even that the bushes smelt unpleasantly. Although I diligently tried to collect as many plants as possible, I succeeded in getting very few; and such wretched-looking little weeds would have better become an arctic than an equatorial flora. The brushwood appears, from a short distance, as leafless as our trees during winter; and it was some time before I discovered that not only almost every plant was now in full leaf, but that the greater number were in flower. The commonest bush is one of the Euphorbiaceae; an acacia and a great odd-looking cactus are the only trees which afford any shade."

—Charles Darwin, *Journal of Researches into the Natural History and Geology of the Countries Visited During the Voyage of H.M.S.* Beagle *Round the World* (1839)

Because the Galápagos Islands seemed to rise magically out of the sea, Spanish sailors in the sixteenth century called them *las encantadas*, the enchanted ones. Herman Melville, who visited in 1841, called them "evilly enchanted," recalling the ancient image of islands as blessed or cursed. These particular ones were certainly not appealing to him. "An archipelago of aridities," he said, "without inhabitant, history, or hope of either in all time to come." However they were described, the effect they had on travelers, with their grotesque land formations and bizarre flora and fauna, was—in both ancient and modern usage—awesome.

"In no world but a fallen one could such lands exist," concluded Melville, referring not only to the geography but also to the whalers and sealers who hunted around there and to the pirates who hid out. But other sailors used the islands, too, and the archipelago became a staging point, or stepping-stone, for travelers through the eastern Pacific. With familiarity, the name Las Encantadas was changed to Galápagos, after the giant land-dwelling tortoises that inhabited the archipelago and whose shell, at the front, seemed to be shaped like a type of Spanish saddle—a *galápago*. But it was the visit of HMS *Beagle* in the fall of 1835, with Charles Darwin aboard, that gave these islands their lasting place in natural history—and in the history of islands.

Darwin, too, was not impressed by his initial encounter with the Galápagos. In a diary entry he was even harsher than in the published account of his voyage on the *Beagle*, comparing the Galápagos "with what we might imagine the cultivated parts of the Infernal regions to be [. . . with] most disgusting clumsy lizards [. . . and] insignificant, ugly little flowers." But his initial disgust was soon transformed into delight when he observed something else on those arid islands on the equator, six hundred miles west of Ecuador: the wondrous diversity of form and function among the plants and animals. He became fascinated by the Galápagos, and they became for him a microcosm of the world, as well as a laboratory of its curiosities and mysteries.

What was especially remarkable, in Darwin's account, was that each of the different islands was "inhabited by a different set of beings. I never dreamed that islands, about fifty or sixty miles apart, and most of them in sight of each other, formed of precisely the same rocks, placed under a quite similar climate, rising to a nearly equal height, would have been differently tenanted." For him, it was not merely about the discovery of new species, but suggested a new understanding of the origin of *all* species.

This challenged the Judeo-Christian belief—shared by many ancient religious traditions—in the creation of each and every species at a single moment in time. Darwin's theory of evolution finally put paid to it, and also to the dating of the beginning of the world at precisely 4004 BCE, a chronology worked out by the Irish clergyman James Ussher in the seventeenth century by counting the generations since Adam and Eve. But in one way Darwin's instinct was not unlike Ussher's—they both counted generations. For Darwin, however, they were the generations in the evolution of a species; and in cataloging them, he intuited the creation of new species according to a chronology that brought the natural world into concert with the patterns and processes of geological time.

The reptiles in particular caught Darwin's attention, with not many species but an extraordinary island individuality among land and water lizards, snakes, and several species of sea turtles. And then there were those saddleback tortoises, which grew to great size and age and on which Darwin frequently rode, getting them moving—he thought they were deaf—by rapping on their shells. These tortoises were indigenous on most of the islands, and their individual island home was even identifiable by distinctive shell markings. And Darwin noticed something else. On the few islands where there were no tortoises, cacti hugged the ground, while on the others the cacti had risen to the occasion and grown into trees, towering high above the tortoises, which would otherwise feed on them. So it seemed that the development of particular plant as well as animal species reflected local conditions. Everything in its proper place, properly.

The fish and other sea life around the Galápagos Islands were also distinctive. Darwin reported that half the seashells were unknown elsewhere—"a wonderful fact," he added, "considering how widely distributed seashells generally are." Interestingly, he identified few insects, and there were hardly any mammals, a situation we now know to be common on distant islands; the exception is bats, several varieties of which are found from the Solomon Islands to Fiji. Of the mammals that Darwin saw on the Galápagos only one mouse, on a single island, seemed likely to be indigenous, though he concluded that the mouse—along with a distinctive species of rat found on another of the islands—might have traveled there on a ship. Regarding insects, Darwin's observation of their scarcity coincided with that of Georg Forster, the naturalist on James Cook's second voyage, who wrote that "the countries of the South Sea contain a considerable variety of animals, though they are confined to a few classes only. No countries in the world produce fewer species of insects than those of the South Sea." Many insects can fly—but not that far.

In Darwin's account, more than half the plants were unique to the Galápagos, and many of them were found on only one or two of its islands, and sometimes in significantly different forms, as with the cacti. The closest kin to the Galápagos plants were in the Americas—and yet, on the Galápagos they were, in Darwin's words, "aboriginal" rather than "immigrant" species, and their environment was different from that of Patagonia and northern Chile, where he had seen similar species. This seemed to him peculiar; we know today, from the evidence of unassuming snails and bugs, that the earliest plant and animal life on the Galápagos originated not to the east, on the continent of South America, but to the west, on what was then a basalt plateau called the Wrangell Terrane, which over the past 140 million years has shifted around to become part of North America (from Alaska to Idaho) but was originally part of the same oceanic plate as the Galápagos. The origin of species was clearly not a simple process.

But Darwin realized that the interrelationship between nature and nurture, which was crucial to the idea of evolution that was taking shape in his mind, was at least simpler on islands than on the mainland, even though the interrelationship between them was more complex *everywhere* than anyone had thought. Nature does what it does wherever species exist, but the dynamics of nurture are more easily isolated in island habitats, with the sea around providing a barrier to travel and transmission and the plants and animals developing in ways that highlight the effects of isolated environmental conditioning, independent of evolutionary developments elsewhere. For evolutionists, this was natural selection in its purest form.

For creationists, on the other hand, the individuality of island species was a feature of creation, with plants and animals either placed there by the creator, their diversity ordained beforehand, or their passage to a specific island decreed within the same grand design that distributed people overland around the world, including to islands. This could have happened only, of course, if islands were simply the leftovers of continental land that had sunk, or because water had risen around them.

And that turned out to be a problem, because with most of the world's islands, it simply never occurred that way. Geology played an important role in demonstrating this by discrediting the "sunken continent–rising seawater" (or "foundered continent") theory of island formation, and in doing so it opened up the discussion of *how*—and in the case of birds and oceangoing reptiles, *why*—certain species ventured to islands in the first place. Today we understand something of the methods of both plant and animal migration to islands, and generally we accept that their arrival would have been by accident—perhaps across land bridges or hopping from one island to another, blown by the wind or carried by the waves or rafting on natural or human craft. By chance, that is, with choice playing a minor role.

But birds are a different story. We know how important a part of island life birds are, and have been for millions of years; but the motives for their travels—where they go, and why (or whether) they stay—are still only partly understood. A few birds might have been blown to the far islands on trade winds; but most would have set out to go traveling. And travel is hard work—the word comes to us from the French *travaille*, with a sense of difficulty. So the questions that surround the movements of birds to and from remote islands seem surprisingly close to those that surround human travel. Does serendipity, a happy accident, a gift of grace, get birds about? Or is it a shrewd sense of purpose, a clever design, a good travel plan? A bit of both, perhaps? Maybe birds have their own stories—but whatever the case, they keep us wondering.

Darwin himself wasn't too interested in a bird's thoughts on the matter. Life is hard work. Motive for him was more or less a synonym for natural instinct. He was interested in what happens when birds get there—wherever "there" is—and most of all, when and how that place becomes home to them. Humans might make a home for themselves on a particular island; but birds have even more choice than humans about where they land up. Darwin was becoming convinced that the interplay between heredity and environment provided an answer to the adaptive changes in plant and animal species that made a particular island their home. And since there was almost always a "control group" of birds on a continent for comparison with their related island varieties, he thought that if he could find out how the interrelationship between nature and nurture affected birds, this would apply to all species and offer insights into the evolution of unique characteristics that are an expression of distinctive "fitnesses"—the fitting into a place—that evolution prescribes.

In Hawaii, later naturalists recognized fitnesses similar to those that Darwin identified, with raspberries, for example, that have no thorns because they evolved in the absence of browsing animals. Indeed, Hawaii has provided a remarkable illustration of island biogeography as well as geology; and one twentieth-century observer noted, "had Darwin called at Hawaii after his historic visit to the Galápagos aboard the *Beagle* in 1835, he would hardly have been able to contain his excitement, for he would have seen examples of evolution which would have made those on the Galápagos pale by comparison. Tiny flies have grown into giants, daisies have turned into shrubs, crickets have gone blind, garden lobelias have evolved into trees, caterpillars have become carnivorous, geese have ceased to fly, and some small birds have even become vampires. There are numerous species of honey-creepers, many brilliantly colored but not all of them eat honey. Some prefer grubs prised from bark and have beaks for the job like woodpeckers, while others have evolved to eat seeds and have tough beaks like finches."

But it was the Galápagos birds that gave Darwin his inspiration, specifically what he called "a most singular group of finches," with peculiarly different beaks "taken and modified for different ends" by the forces of evolution: finches with long beaks, to catch insects; and finches with short beaks, to crack nuts. The ornithologist Roger Tory Peterson would later call them "the finches that shook the world," in tribute to the seismic effect of the theory of evolution which Darwin began to formulate on these enchanted isles.

Typically for islands, all kinds of birds were to be found on the Galápagos. There were waders and waterbirds aplenty, though because they are habitual wanderers, fewer of them were unique to the archipelago. But twenty-five kinds of land birds—including hawks, owls, wrens, flycatchers, doves, swallows, thrush, and finches—were found nowhere else. Flightless birds, like the cormorants on the Galápagos, particularly intrigued Darwin. In the storyline he proposed, they would have evolved there in the absence of natural predators, and then some of them would have been wiped out when humans or animals—such as certain kinds of snakes and toads—arrived and made their home inhospitable. This is what happened to the great auks on the islands of the North Atlantic (obliterated by seafaring hunters), the moa in New Zealand (extinguished when the Maori came), the dodo on Mauritius (which died off when humans destroyed its habitat, and when imported animals such as dogs and pigs and cats ate its eggs), and the ibis in Jamaica (probably done in by snakes, monkeys, and other birds). Fortunately, some flightless species have survived, such as the wood hen on Lord Howe Island in the Tasman Sea, the Hawaiian goose, and the kiwi in New Zealand—the latter two now celebrated as the state bird and the national bird of their respective island homes.

For people who believed that all species were created uniquely, at one time rather than evolving over time, the flightless birds would have walked to, or already lived in, those parts of the mainland that transformed into islands when the land sank or the water rose. In their view, such islands were the remnants of various foundered continents, like those fanciful sunken continents or stepping-stone isles that were once said to account for human settlement in Polynesia. This explanation had scientific as well as religious advocates in Darwin's time, and one of his most important legacies was to draw on geological evidence and demonstrate its impossibility. In doing so, he brought the origin of islands and the origin of species into the same story. And birds were a central part of this.

Remarkably, it was not the wildness of certain island birds that puzzled him—it was their tameness. One of the most obvious survival mechanisms for a prey animal is an instinctive fear of its enemies and a set of defenses—for instance, running and kicking and biting for a horse—that will help ward off trouble. Birds typically keep their distance, which is easier when you can fly, and is why the flightless birds became so vulnerable. The tameness of the wild birds that he had seen earlier on the Falklands, and now observed again on the Galápagos, led Darwin to one of his most important conclusions about the inheritance of acquired characteristics. "In regard to the wildness of birds toward man," he wrote, "there is no way of accounting for it, except as an inherited habit: comparatively few young birds, in any one year, have been injured by man in England, yet almost all, even nestlings, are afraid of him; many individuals, on the other hand, both at the Galápagos and at the Falklands, have been pursued and injured by man, but yet have not learned a salutary dread of him. We may infer from these facts, what havoc the introduction of any new beast of prey must cause in a country, before the instincts of the indigenous inhabitants have become adapted to the stranger's craft or power."

To this day, as one astute observer has remarked, it is a Galápagos paradox that its wild animals are tame and its once tame animals—the sheep and goats and cats and pigs left by sailors over the centuries—are wild. And as if to take up this tale, far beyond the Galápagos the categories of the wild and the tame, or the barbaric and the civilized, have been conflated or confused or sometimes celebrated by generations of island storytellers, with humans being portrayed as subject to the same evolutionary forces as every other species.

As the theory of evolution was taken up by Darwin's colleagues and successors, the role of island species became central. Over time, exemplary island forms included that singular group of finches in the Galápagos—whose long beaks and short beaks were acquired by genetic mutation and inherited over time—and those Hawaiian honeycreepers, with beaks likewise adapted to feeding on different foods on different islands; the ants of Melanesia, which, in the twentieth century, inspired the biologist Edward O. Wilson to develop his theory of sociobiology; and the land snails of southwest Pacific islands, whose rich variety and wide dispersal continue to intrigue biologists. All of these animals were flourishing in unique island environments "controlled" by the sea around. Plants provided other illustrations, from the ground-hugging and the high-rising cacti on the Galápagos to lobelia on the island of Oahu in the Hawaiian archipelago, with broad leaves in the tropical lowlands to catch light, thick leaves to retain moisture on the windward mountain slopes, and shiny narrow leaves on the leeward side to counter the intense light.

As Darwin's ideas spread, a wide range of environmental conditioning also became apparent in the fossil record and in cave paintings and other images drawn by humans millennia ago. For instance, it is now known from such evidence that during the last Ice Age certain mammal species on islands in the Mediterranean—including elephants and hippos, which we normally associate with mainland Africa—experienced dwarfing over time as a different kind of environmental nurturing took place, just as miniature wooly mammoths evolved on Wrangel Island in the Russian Arctic. (For most animal species, getting smaller takes less time, in evolutionary terms, than getting larger, because there aren't the same "structural engineering" challenges.) And in a nice reminder of how mythmakers and scientists have often sung in chorus, the remains of small adult "humans"—about three feet tall—who lived tens of thousands of years ago were found on the island of Flores, east of Bali, in 2003. Indonesian hobbits, so to speak, courtesy of J. R. R. Tolkien.

The Chinese word for an island reminds us that birds have been associated with islands for a long, long time. Describing an island called Funk, off the coast of Newfoundland (great auks once lived there, and it was first named the Island of Birds when it appeared on one of the earliest maps of the western Atlantic, in 1504), the nature writer Franklin Russell caught the mysterious affinity of islands and birds. "Inside the island, a million birds sleep. The forest and grass conceal them and drip with the moisture of a night-borne mist [. . .] When I am drowsy and ready to sleep in my apartment in Manhattan, I think of the island. It rests in a part of the mind where its smells, colors, and sounds, its hoary rocks and hordes of creatures, can be transformed instantly into an image in my eye. I see it under a cool sun as its day ends and night flows from the sea. But the island does not sleep; instead, it awakens."

However romantic this may sound, the current name of the island, Funk, comes from its smell. Birds fill out the sky there in certain seasons, sometimes more than a million in a few square miles, covering everything with their droppings. Before chemical fertilizers, such droppings—politely known as guano—were so valuable that the United States, when it learned of the rich deposits on various Pacific islands, passed what became known as the Guano Act (1856). This authorized Americans to take possession of any island that was not occupied or claimed by other governments and to defend such interests by military force. It was under the Guano Act that the last substantial island to be identified in the Pacific by Western explorers was annexed by the United States in 1859, and was then called Midway Island.

Harvesting guano wasn't just an American enterprise. A series of islands off the coast of Peru, the best-known being the Ballestas, are home to over one hundred and fifty species of birds, including pelicans, cormorants, and the Humboldt penguin, named after the current that flows north up the Pacific coast, carrying cool water from the Antarctic and past the Juan Fernández Islands. But the current also creates a curious weather pattern on the west coast of South America, with very little rain, and therefore little vegetation, on the islands offshore. As a result, the bird droppings that accumulate there are baked dry, preserving the nitrates that make fertilizer valuable. For centuries, guano from these islands had been much prized, and during the 1800s, the sale of guano to Europe provided Peru with its most important source of revenue.

Almost all of the world's islands provide either a home settlement or a wanderer's rest for birds, though not all of them are covered with droppings. Islands called "Bird Island," especially if all local names for particular birds were included, must number in the thousands. There are islands named for the raven and the pigeon and the dove and the canary, for the eagle and the osprey and the seagull and the swan, for the albatross and the petrel and the puffin and the duck—and for birds whose names are held safe in the languages of the peoples of the Pacific and Indian and Atlantic and Arctic oceans.

"Bound is the boatless man," say the people of the Faroe Islands. Birds, except the ones that can't fly, are never bound to one place, except by—well, by what exactly? Why do some birds remain during the harshest weather and the leanest times, watching others—often less capable of long flights—pass over on their way to fairer climates? Although we can come up with reasons, from old habit to new habitat, and from breeding and nesting environments to staple foods (such as insects that don't survive certain climates), we can't always explain why certain birds seem to prefer one particular island to another (or why they choose one over the other as a place to stay awhile). Inertia, of course, cannot be discounted; it is one of the great forces in geological, geographical, biological, and human history, and it is an inevitable feature of island life—but it only explains so much. Whatever answers we propose, none of them cover even all the birds of a particular species. Starlings, which are notable migrators, stay on the Faroe Islands all year—while oystercatchers, which often stay in one place, migrate there every year. There is evidence that some species (monarch flycatchers, for instance) "backtrack" their island-hopping routes and recolonize the lands from which they originated, once again for reasons we wonder about. Why do they? And why don't others?

"Speed bonnie boat, like a bird on the wing [. . .] Over the sea to Skye" begins the "Skye Boat Song," written in the middle of the nineteenth century to celebrate the flight of Bonnie Prince Charlie to the Isle of Skye after his defeat at the battle of Culloden in 1746. For countless generations of poets and songwriters, birds have represented the spirit of freedom, the envy of all those who remain wingless or boatless and bound. Just like people, though, some birds prefer the nooks and crannies of home, while others fancy the open sky (and many like both, but at different times). And all birds, like all people, need a bit of land, no matter how modest, to alight on every once in a while.

Even islands named for one particular bird offer home or haven to many other kinds. There are birds that perch and sing (they are called the passerines, and they make up over half of the birds on the planet) while others walk and talk. Some remain in one place all their lives, and others travel from island to island, or right across the ocean (like modern humans on vacation) to catch the seasons, often non-stop, with a few (like the swift) even sleeping en route. Some birds appreciate the land, but others (called pelagic) choose to stay at sea for long periods. Still others settle on shore, where land and sea are in constant conversation. Many enjoy company, while turnstones, for example, are so territorial that they will chase any odd bird that bothers them, sometimes up to a hundred miles.

On ocean islands there are herring gulls, crying in the breeze and strutting on the rocks; ducks paddling about, upending themselves to catch fish just under the surface; and cormorants, sometimes called "sea crows," diving down fifty feet for food. There are long-living sooty shearwaters, known in many places as puffins (but as hagdowns in Newfoundland), which stay at sea all winter and may live for over half a century; and black-legged kittiwakes that build their nests on the cliffs in summer and return to the water in winter. Murres, from the auk family, are common in the North Atlantic and North Pacific, some nesting as permanent residents while others migrate from Arctic islands to temperate climates in the winter. They fly fast—up to fifty miles per hour—but are much more agile underwater, where they use their wings for propulsion and can dive to depths of six hundred feet. Petrels and albatrosses are the constant companions of sailors all over the world, often living mostly at sea for years, smelling food from miles away and drinking the saltwater that sailors can't stomach. Some say albatrosses bring bad luck—as in Coleridge's "Rime of the Ancient Mariner"—while others say that albatrosses are birds of good omen. But everyone agrees that they warn of a change in the weather. They are still abundant in the Pacific, though feather merchants almost wiped out the short-tailed ones (also called Steller's albatross), taking ten million in a few years at the end of the nineteenth century. Over the coast of some islands eagles float on currents of air as gracefully as seals under water, landing in the ancient trees that guard the shore. Ravens scavenge wool from island sheep to line their nests, and screech and show off by flying upside down like trick pilots at an air show and bothering the eagles—who pretend they don't notice.

We cannot imagine an island, or indeed a world, without birds; and the British islander John Keats, in his haunting poem "La belle dame sans merci," caught the ultimate image of desolation. "The sedge has withered from the lake, / And no birds sing."

To qualify as one of the Thousand Islands in the St. Lawrence River bordering Canada and the United States, an island should be over one square foot in size, be above water three hundred sixty-five days a year, and support at least one tree. But not all island trees appeal to everyone. When Magellan made his way into the Pacific, his chronicler Antonio Pigafetta described the first islands they saw as "uninhabited [. . .] we saw no other things than birds and trees, and for this reason we called them the Unfortunate Isles." It's all a matter of perspective.

Mangroves are at home on the muddy shorelines of many subtropical and tropical islands (and mainlands), and even though they are not very hospitable to humans, they protect and promote an island by extending its reach into the ocean (up to a hundred feet per year along some of the Borneo coastline, for example). Rain forests are found on both temperate and tropical islands, often giving way to upland trees and cloud forests. The giant cedars and spruces and fir trees on the islands of the northwest coast of North America tower in the imaginations of Aboriginal peoples who have been there since time immemorial. Their stories tell of people who were on the islands of Haida Gwaii even before there were the cedars from which they later made canoes. They settled on the rocky shores, ate what the sea and the sky provided, clothed themselves in fish skin and bird feathers, and waited for the trees that would make their island home and their seafaring possible.

The windward (or wetter) slopes of islands typically nourish luxuriant growth, while on the leeward side there may be shrubs and scrub timber, though sometimes that is where the most valuable trees grow—like the sandalwood, which became a major trading commodity in the late eighteenth century, with stands discovered on Fiji, Hawaii, and the Marquesas, and later on New Caledonia and Vanuatu. On islands where lower slopes have been cleared of trees by humans, secondary growth emerges to diversify island ecologies and continue the cycle of island life. The overbearing banyan was originally brought from the South Asian mainland and now flourishes on various Pacific islands, sometimes taking up several acres with its aerial roots branching off from the main trunk. The fictional Robinson Crusoe made his home in a banyan tree.

Despite our fancy of an Edenic originality on the tropical Pacific islands, it was probably Polynesian settlers who brought many of the trees and plants we associate with those islands, including coconut palms, chestnuts, breadfruit, bananas, taro, sweet potato, arrowroot, and yams. As one chronicler puts it, "the pre-European islanders may indeed have lived in a Garden of Eden, but the garden was planted and tended by man," sustained by climate and fertile land and maintained by the imposing barrier of water surrounding the distant islands. On the other hand, the grasslands long associated with human settlement on some of the Pacific islands are now thought to be as much the product of naturally occurring fires and rainfall fluctuation as of human practices (though some savannah lands would have been created by the cutting of trees for clearing and construction). When the Portuguese first arrived on St. Helena in 1502, they remarked on the lush growth of trees hanging over the cliffs. Later, these trees had disappeared, and settlers were blamed; in 1771, the ubiquitous Joseph Banks said that they had made "a desert out of paradise." But it might not have been entirely the settlers' fault, for the ground may have been naturally cut out from under the trees on the receding shoreline. The idea of islands as places of natural environmental balance needs some qualification—and so does the notion that everything bad that happens on islands is caused by people. Even the standard story of what has been called Easter Island's "environmental suicide," the poster child of catastrophic ecology and the convenient explanation for the near-disappearance of its human population, is now being challenged on a number of grounds. The destruction of trees on the island, and the extermination of much of its other flora and fauna, was almost certainly the result of both environmental and human factors—both of them "natural," in any language. Haunted by the monumental statues that remain, some would say that the supernatural, too, may have played a part. And in line with the insights of Darwin and his colleagues about the idiosyncrasies of island flora and fauna, it is becoming clear that climate change and water-level variation can have significantly different impacts even on islands in the same region, just as the health of codfish stocks can vary from one small harbor, or outport, to another on a single island such as Newfoundland.

Nevertheless, the changes in island flora and fauna that have been introduced by humans have occasionally had disastrous consequences, even as the people flourished. One of the most surprising of these was the result of an admirable human instinct—to help others. Sheep, goats, pigs, and cattle left on islands as food for passing ships not only kept castaways like Alexander Selkirk alive but also grazed on or rooted up the undergrowth, exposing the soil to erosion and stunting the regeneration of shrubs and trees—until everything was browsed into oblivion. Grazing and browsing animals, very seldom native to islands and never to the ones far offshore, brought about landscape changes that damaged or destroyed native species. The effects on the island of St. Helena were studied by the British naturalist Alfred Russel Wallace, author of the classic *Island Life* (1880), at a time when goats—astonishingly—were worth more on the island than the ebony trees that were being cut down there. On many islands, citrus trees planted by early European explorers to provide relief from scurvy for future seafarers seem to have been a relatively benign intruder; but not so the livestock given to Tahitian chiefs by Wallis, Bougainville, and Cook. And rabbits have a particularly sorry history (though not on Tahiti, where there were plenty of omnivorous dogs and pigs to eat them), often transforming island environments as well as continental "islands" like Australia. Some say that such animals do not "belong" there, and that they bring dangerous knowledge to a wisely conceived, well-balanced garden of Eden. The idea that particular islands are the proper place for certain species but not for others reinforces the sense of what might be called "evolutionary propriety"—almost as though it had all been ordained, or intelligently designed.

When Darwin turned from the geological to the natural history of islands, it was because he recognized—like his friend Wallace—that they would provide a set of case studies for the dynamics of both organic development and environmental influence. Wallace's intuition about the inseparability of the physical and natural history of islands was similar to Darwin's; but for a time they disagreed, not about the origin of species but about the origin of islands. Wallace had done much of his early work (from 1854 to 1862) on the coastal islands of Indonesia, where the islands rise from a continental shelf; and it was in such an environment that he developed his ideas about the geographical distribution of plants and animals, and about islands as the leftover product of continents that had sunk—the foundered continents once again. Darwin, on the other hand, began out in the ocean (home to most of the world's islands), where the continental origin of islands was dubious.

But since marine fossils and sedimentary rocks on all the continents confirmed that they had once been under water, the proponents of foundered continents assumed that the ocean floor must once, in the distant past, have been land before it had sunk. This seemed a reasonable scientific explanation; it had an elegant symmetry to it, which scientists like, and it was also appealing to religious people because it allowed for a single act of creation. Aware of the rise and fall of sea levels over time, and with little knowledge of the ocean floor (and no knowledge of plate tectonics), many people from a wide range of backgrounds held onto the idea that when the sea had been low, people and animals had wandered, consciously or unconsciously carrying the seeds of plants with them; and when it had been high, they had settled down, wherever they happened to be. Even Darwin's friend and scientific ally, the distinguished botanist Joseph D. Hooker, argued that the diversity of nonhuman life-forms on islands could not be explained by random dispersal across the seas, while the problem went away if the islands were simply the peaks of sunken continents.

To Darwin it became clear that foundered continents could not explain the Pacific Ocean islands. Instead, as we have seen, he believed that physical forces on the ocean floor were the cause. If islands like the Galápagos were simply left over from continents that had been overwhelmed by water, then their flora and fauna would be nurtured primarily by continental conditions rather than by the specific conditions of particular islands. His observations of the diversity of island species and the identification of unique new species in island environments made that account unreasonable, and geologists soon made it untenable. In geological terms, if sea islands were in fact the peaks of foundered continents, they should show evidence of the same sedimentary rocks as the Himalayas, for example, or granite like America's Sierra Nevada. Most islands did not, and Darwin was encouraged in his conclusion by the volcanic (basalt and coral limestone) composition of many islands in the ocean, from Hawaii and Tahiti to St. Helena and the Azores. When Wallace wrote to Darwin about his own work in 1858 (the year before *On the Origin of Species* came out), Darwin wrote back agreeing with everything that Wallace proposed (which closely mirrored Darwin's own ideas)—except for the foundered-continent idea. In short order, Wallace accepted Darwin's arguments, which became fundamental to evolutionary theory. It may have been the naturalists that threw the fat into the fire in the fight between science and religion, but it was the geologists who laid the fire.

Darwin insisted that the rising—and eventually the falling—of land in the middle of the sea was not just the most plausible explanation for the origin of islands but in fact the *only* way of accounting for the eccentric ecology of island life. This was the foundation of his theory of evolution. Darwin had read his Bible, the book of God, with its story of the creation of the world when the waters under heaven were gathered together in one place and dry land appeared, and then of its destruction by flood. But he was beginning to read another book, the book of Nature, and for Darwin it began as island reading. That was where he realized that the "infinite variety" celebrated in the hymns he had learned as a child and sung as he prepared for a career as a clergyman was not so much an image of God's ingenuity but evidence of independent evolution on different islands—only a few of which turned out to be the remnants of continents—out in the ocean.

"Survival of the fittest" was not Darwin's phrase (the philosopher and scientist Herbert Spencer coined it) but "natural selection" was, and it became the hallmark of evolutionary theory, inviting us to imagine a selector—an intelligent designer, if you will—at the same time as it eliminated such a designer, leaving all to chance in the form of heredity and the environment. Darwin was well aware of this contradiction at the heart of his theory, and while he strenuously denied the implication of purposefulness, it continued to shadow the scene. And perhaps it always will, for something in us seems to want to maintain a balance between determinism and free will, or chance and choice. Which may be why we sometimes give nature an almost supernatural capacity to "get back at us" if we abuse it, and use a word with ancient religious associations—the word "pollution"—to describe our desecration of the environment. And with what we now know about genetic inheritance, Oscar Wilde's description of heredity as "the last of the Fates, and the most terrible" has become more plausible than ever.

Cultivated crops have had a significant influence on island flora and fauna from the early days of human habitation. Some of the crops are the product of island capacities, while others come from cultural preferences. And almost all find an important place in myths of origin that describe the beginnings of life itself, or that of islands.

On Pacific islands, the coconut, breadfruit, and the taro became the staple crops, along with yams, sweet potatoes, pandanus, and bananas. Being salt tolerant, coconut could have taken root on islands before human habitation, floating there across the water. The limited cultivation of rice and other cereals, notable in Polynesia, may have been the consequence of soil or climate conditions; some say that the Polynesians did indeed bring rice to the islands but replaced it when they discovered the advantages of the other high-yielding and healthful plants that would grow in the alternately dry and wet environments (on the leeward and windward sides) of mountainous tropical islands. On the coastal islands of Southeast Asia, on the other hand, rice has been an important food for a long time, and it figures in many stories, such as a Dayak myth in Borneo that tells of rice originating as a drop of milk that fell from the breast of a celestial god.

Stories about the origin of the most important food plants are very common in island (as in mainland) communities, with the number of myths often being proportional to the importance of the plant in the local diet. Taro is the subject of stories connected to Papa and Wakea, the mythical ancestors of the Hawaiian people and identified throughout the South Pacific as the goddess of the earth (or the underworld) and the god of heaven (or light), respectively. On Tahiti, breadfruit provides the more widely invoked mythology. It is often described as originating from the parts of a human, sometimes associated with self-sacrifice to save others. In Hawaiian stories, the breadfruit is brought to the island from elsewhere, variously by accident or design. One Hawaiian myth identifies the upright breadfruit tree as male and the spreading tree with overhanging branches as female, each with lineages that bring together the human and the divine. There is a story from Rarotonga in the Cook Islands that tells of a god-like hero having once lived in a place where the people ate rice, but he refused. When he died, his wife buried him in their yard and a breadfruit tree rose from his grave.

Eventually, breadfruit became associated so closely with the rich food sources found on the Pacific islands that Captain Bligh returned to Tahiti after having survived the mutiny on the *Bounty* to secure stock for Jamaica, where it was used to feed enslaved Africans working the sugar plantations. A series of hurricanes there in the 1780s, which rooted up the plantain walks, was followed by drought and a famine in which as many as fifteen thousand Africans may have starved—at a time when foodstuffs from the American mainland were effectively unavailable to British Jamaicans in the aftermath of the American Revolutionary War.

No tree is more closely associated with tropical islands—and with islands of the imagination, at least for those of us who live in temperate climates—than the coconut palm. It seems to have generated the most stories in the Pacific, many of them associated with the trickster hero Maui. One story that has wide currency tells of an undersea goddess who flees her eel husband and asks Maui for protection. He kills the eel-god, buries his severed head, and from it sprouts the first coconut tree. The goddess then releases her calabash of food, which become the moon and the stars. The association of coconuts with a decapitated eel is also part of a set of tales from Tahiti, and many myths throughout the region identify Tahiti as the island where the coconut originally came from. The connection between a social or spiritual taboo and the substance or shape of a plant is common in many origin myths; the height of the coconut palm is explained in one story from Hawaii as being the result of a lowly mortal trying to pick coconuts belonging to the god Kane—but as the man climbed, the tree grew higher and higher and he could not reach the coconuts.

A story from New Caledonia tells of a man fleeing to a desert island with no food or water. A god appears and instructs the man to bury a stone, which becomes a coconut tree. In a remarkable way, this creation myth mirrors our contemporary cartoon image of a castaway sitting under a tropical palm; and this is much more than a cartoon illustrator's invention, for there is a wide range of stories across the Pacific in which the coconut figures as a signature of island well-being.

In luxurious hotels, the original palm of the palm court—an island of tranquility for the affluent urbanite—was the kentia (or thatch) palm from Lord Howe Island. For Europeans, the palm may also present a complex legacy of the classical association of palms, including triumph over one's enemies (as on Palm Sunday) and victory in a game (as celebrated by Romans with a palm branch). Even the name Phoenicia, where some of the greatest island sailors of ancient times came from, is said to mean "land of palms."

There is an island in the North Atlantic, between Ireland and Iceland, called by one of its recent chroniclers "the smallest isolated rock, or the most isolated small rock" in the oceans of the world. Not surprisingly, its name is Rockall, though an eighteenth-century account of the western islands of Scotland refers to it as Rocabarra, identified in Gaelic myth as a rock that appears three times, the last one being at the end of the world. Geologically, Rockall is the core of a long-extinct volcano; it is less than a hundred feet across, almost as high as it is wide, and two or three hundred miles from any other land. It has no trees, and indeed supports few plants and animals, though there is some vegetation, with several species of seaweed and periwinkles, along with mollusks and worms and mites and a familiar array of birds, nesting for awhile or just passing through—petrels, puffins and shearwaters, gulls, gannets, and guillemots, the occasional tern and, long ago, the great auk. The winds and fogs of the

North Atlantic make it a dangerous outcrop, impossible even to land on in most of the weather that graces that part of the world, and it has seen more than its share of wrecks.

But you never can tell with islands, and the waters round Rockall hide a cold-water coral reef of extraordinary variety, with sea anemones and sea spiders and sea urchins and many as yet unidentified species. Rockall also has an interesting political history (a natural history of sorts), and it may even have some future economic prospects. In 1588, the remains of the Spanish Armada, returning home in defeat and driven off the west coast of Ireland by stormy weather, had a rendezvous at Rockall. In 1955, a helicopter landed a small group from a British navy survey ship on Rockall to claim possession of the island in the name of Queen Elizabeth, which made it the last territorial expansion of the British Empire. And nowadays the island is coveted for its location on the continental shelf that reaches out into the North Atlantic, with potentially significant oil and gas reserves. Greenpeace protesters set up camp there in the 1990s to try to establish international sovereignty and fend off underwater development. But with those undersea resources in mind, its jurisdiction is now disputed by the United Kingdom, Iceland, Ireland, and Denmark on behalf of the Faroe Islands.

The Faroe Islands themselves are part of a mid-ocean ridge of volcanic basalt which once rose above sea level and formed a plateau that covered nearly three quarters of a million square miles. Now reduced by erosion to eighteen separate islands lying halfway between Scotland and Iceland in the North Atlantic, the Faroes have both cultural and geological ties to the islands of Shetland, Orkney, and the Hebrides. Islands chain-linked to each other in archipelagoes, like these and other islands in the North Atlantic, often provide resting places not only for birds but also for travelers by boat—like the old overland post-houses where you could rest your horses, have a meal, and get some sleep—reminding us that the sea has connected as well as divided people for millennia. For a long time, stepping-stone islands made long voyages possible for humans, almost certainly prompting many of the early migrations that peopled the islands of Indonesia and the Philippines as well as the Mediterranean, and encouraging seafarers from northern and southern Europe to travel to the Canaries and Azores. In the North Pacific, the remains of Beringia, the ancient land bridge, is now reduced to the two Diomede Islands—one Russian, the other American—in the Bering Strait, but still connecting Asia and the Americas in an Inuit heritage of walrus hunting that has lasted for millennia. Caribbean stepping-stone islands were well worn by the Arawak migrating from the American mainland; and throughout the Arctic, ice islands have provided resting points for seals and polar bears, and staging grounds for those who have hunted them.

The Faroes were originally inhabited by wandering saints and Irish settlers who, beginning in the sixth century CE, rowed and sailed there, bringing oats and the sheep that gave the islands their name (the word *faroe* means sheep). By the eighth century, the characteristic beehive huts built by monks on many of the North Atlantic islands provided temporary shelter (alongside fresh vegetables from their island gardens) to seafaring travelers. One of the Faroes was named Paradise Island of Birds by St. Brendan (also known as Brendan the Navigator), a sixth-century visitor on the lookout for his blessed isles, whose legendary voyages in the Atlantic became the basis of the medieval bestseller *The Voyage of Saint Brendan.* Appropriately, the Faroes are now famous for their birds. And for their sheep. (The definition of an island in a Scottish census taken in 1861 required that an island must support at least one person and one sheep. In this case, trees were not a requirement.)

When the Vikings arrived in the ninth century, they turned from a tough life on the land to the abundant sea, and to a diet of fish, seals, birds, and their eggs. Indeed, the eggs of birds along with the variety of plants and animals on beaches and in tidal pools have fed folks on the islands of the world for millennia. On the Faroes, the harvesting of puffins and guillemots, once a major source of protein, has been on the decline, though annual catches of over two hundred thousand puffins and one hundred forty thousand guillemots were still recorded in the 1940s. The white and black Eurasian oystercatcher (called *tjaldur*) is the national bird of the Faroe Islands, and the islanders celebrate its arrival—as regular as clockwork—in early March every year, signaling the beginning of spring and trouble for shellfish. Six months later, almost to the day, the oystercatchers leave on their annual migration to warmer shores around the Mediterranean and North Africa. Flycatchers visit the Faroes in the summer, after having wintered in West Africa. By far the largest numbers of birds on the Faroes are the seabirds, with over two million breeding pairs. Some of them, like the fulmar, feed from the surface of the sea, while puffins dive as deep as five hundred feet. Great skuas and arctic skuas are also found there, along with gannets and gulls, kittiwakes and razorbills and that remarkable long-distance flyer, the arctic tern.

Like birds, seals and sea lions are archetypal island dwellers. Grey seals are abundant on the Faroes, and have long been a staple of the diet for the humans living there. The Juan Fernández Islands in the South Pacific are home to a species of fur seal that exists nowhere else, finding refuge on hard-to-reach shorelines or rock outcrops, thus relatively safe both from their land-based enemies and from the killer whales who swashbuckle in the sea. When the seal pups are born, the rockeries serve as their nurseries, where they lollygag about while the elders look on with parental concern, making superior-sounding noises before taking the youngsters out into the coves and inlets to teach them to hunt for food and watch for trouble.

On islands, seals are relatively safe from the predator animals that hunt on the mainland. Except for humans. We have probably hunted seals and sea otters and sea lions from time immemorial, first of all taking advantage of their clumsiness on land, and then chasing them in the water. But it was the worldwide commercial hunt for their skins and their oil that proved fatal to many seal and sea otter populations. The Russians had been trading in skins and oil in the North Pacific around the Kamchatka Peninsula since the early eighteenth century, and trade also developed further south, in the 1750s, around the islands of Hawaii; but following the spectacularly lucrative sale of sea otter skins by sailors on James Cook's third voyage, word spread quickly and the hunt—a seafaring gold rush—was on. Seals were killed on islands around the world during the eighteenth and nineteenth centuries, hunted both on land and at sea. Many seal rookeries had been completely destroyed in the North Pacific by the end of the eighteenth century, and there was a decline in populations almost everywhere. As sealers sailed further into the southern oceans, the same thing happened. Each island discovery was like a find of gold or silver, its location kept secret until the vein was exhausted. Some seal and otter populations were wiped out; and others went into hiding. When sealers came to the Juan Fernández Islands, the seals there seemed doomed, with each sealer skinning up to a seal a minute through the day, and the pelts sold to China and Europe. By the middle of the nineteenth century, it was said that there were none left; but a few survived, probably in sea caves, and there are now ten thousand living on the islands of the archipelago.

The trade in sealskins, like that of beaver pelts and other land-based animal furs in North America and Asia, produced extraordinary harvests and extraordinary profits. Seal hunting still continues, with some poaching and the accidental—though mostly avoidable—entanglement of seals in commercial fishnets, but much has been done to protect surviving island species on the Juan Fernández and Galápagos archipelagoes, on the islands of the North Pacific, and in the southern seas around Australia, New Zealand, South America, and southern Africa. The infamous Robben Island off South Africa (where Nelson Mandela was imprisoned) was a center of seal hunting from the seventeenth century—the word *robben* means seal in Afrikaans—but for the past century there have been no seals there.

The story lives on in the Newfoundland seal hunt, the epitome of barbaric savagery for some—and subsistence harvesting for others. The seal population there remains reasonably healthy, with some arguing that a regular culling of the herd is necessary; and no one who has been out on a hunt would question the dangers and harsh demands of it on the ice floes of the North Atlantic, part of a tradition that goes back hundreds of years and was deadly for some of the hunters, marooned for a couple of months on islands in cold and wet weather, living in caves or under overturned boats, hunting at night over slippery rocks or stranded on ice floes. But the seals got the worst of it.

It was oil as much as skin and fur that drove the trade—seal oil and whale oil (also called train oil, after an archaic word meaning "squeezed"). By the late 1700s, European and North American whalers were exploring and exploiting the resources of the Central and South Pacific, and later the North Pacific, in an enterprise that was dangerous for the whalers and horrific for the dying whales, who struggled—sometimes for days—while hooked with a harpoon. Many of the Polynesian islanders who were caught up in the hunt were used almost as cruelly, though whaling did not play favorites, as Herman Melville made clear in his novel *Moby Dick* (1851). And whaling had been part of maritime life long before recorded history, with whale bones and harpoons discovered in Stone Age middens from six or seven thousand years ago. It was Basque whalers, from the same communities that fished cod around Newfoundland in the centuries before Columbus, who developed the techniques of harpoon and line as early as the eighth century CE. These techniques became the staple of Pacific whaling a millennium later, when the eighteenth- and nineteenth-century whale trade was centered around ports like Papeete in Tahiti, the Bay of Islands in New Zealand, and Honolulu. The trade was immensely profitable through most of the nineteenth century, though after the first petroleum wells were drilled in the 1850s, whaling began to lose its virtual monopoly on oil.

Amphibians and other reptiles are associated with all islands, though their distribution depends on a number of factors, primarily proximity to large landmasses and habitat. Distance determines which reptiles make it to an island on their own; and because food supply is such a major influence on the development of any species, it might be assumed that the more food an island offers, the bigger the animal—and vice versa. But there is a surprise here, for it is not simply genial weather and a generous table that determine size. If food is unpredictable, reptiles require a large body mass to store fat to keep them going through lean times; therefore very large reptiles are often found on very inhospitable islands. There are the giant Galápagos tortoises, for example, in an arid island landscape. The enormous Komodo dragons—the world's largest lizard, up to ten feet long and weighing over three hundred pounds—are found on only a handful of southern Indonesian volcanic islands, in fairly harsh desert conditions (with rain only in monsoon season, when flooding is typical). And an unusually large gecko lives on a small island north of Mauritius, which has a scattering of palm trees but a habitat severely inhibited by local cyclones. Small reptiles are another story, with little geckos and burrowing snakes on the Caribbean islands and tiny chameleons on Madagascar, their size allowing them to eat the even tinier prey common in their island environments. Not to mention some very strange prehistoric island mutations, such as an extinct dwarf goat with significant reptilian characteristics identified in fossils that have been found on the island of Majorca.

We turn to islands when we think of strange species, just as Darwin did. Australia has enough odd creatures to give it status as an island, even though its geology is continental; but the diversity of the island of Madagascar, sometimes called the eighth continent because of its extraordinary variety of plants and animals (and perhaps, too, because it is one of the largest islands in the world) is equally exceptional. Ninety-five percent of its more than 300 species of reptiles, over half of its 200 species of birds, and over 80 percent of its flora and fauna—including 95 percent of its trees and large shrubs—are found nowhere else; and a full 5 percent of the world's species of plants and animals are represented on the island, many of them endemic. All the land mammal species are unique to the island, most notably the lemur—which first came there from Africa on rafts of vegetation over fifty million years ago. Madagascar also has the second-highest number of primate species in the world, after Brazil.

And there are many other islands that have notable flora and fauna. Some have a limited variety, such as Ascension in the South Atlantic, about midway between Angola and Brazil, which is home to a modest number of indigenous plants, fish, snails, and worms, as well as being an important nesting site for green turtles and a breeding site for terns and tropicbirds as well as noddies and boobies, with the Ascension frigate bird endemic to the island. Much more spectacular is the archipelago of Socotra, east of the Horn of Africa and about two hundred miles south of Yemen. Socotra (the name comes from a Sanskrit word meaning bliss) is not volcanic and was once part of continental land, although it has long been isolated, with a geography as forbidding as that of the Galápagos. One third of its over eight hundred plant species, many of them very strange, are indigenous; and of about forty species of birds there, ten are unique to Socotra.

Way down in the South Atlantic, the Falklands (known as the Malvinas to the Argentine people on the mainland, three hundred miles to the west) must hold some kind of record for bird populations, colonial settlements, and international contention, so much so that one commentator has called them "the islands of birds, sheep and wars." (One of these islands is of course called Bird Island.) Oystercatchers flourish there, including the Magellanic oystercatcher, native to that part of the world, as well as a bird of prey called the caracara, or Johnny Rook. Darwin, who visited the Falklands during his *Beagle* voyage, didn't much like the Johnny Rook, and described it as "exceedingly numerous, very mischievous and inquisitive, quarrelsome and passionate." And astonishingly tame, which intrigued him, setting the stage for his later speculations on the Galápagos. Eventually Johnny Rooks got into trouble with settlers by bothering the imported sheep that became the staple of island life; and in due course they were hunted for bounty. Now Johnny Rooks are under threat of extinction, with only about five hundred breeding pairs left. But nobody seems to have told the birds, for they continue to behave in the same old way, rivaling the raven as tricksters and thieves. The Falklands have many other birds, too, including a wide variety of penguins and petrels, sandpipers and snipes, gulls and geese, cormorants and albatrosses, finches and meadowlarks and the inquisitive tussock bird, found only on the islands in that region. And shearwaters, up to twenty thousand of them, breeding there in the South Atlantic and then packing up and heading to the North Atlantic islands for the summer.

Like many islands, the Falklands are home to sea lions and seals, specifically fur seals and elephant seals, the latter so called not because of their size but because of their noses, which offer a good imitation of an elephant's trunk. Elephant seals spend most of their lives in the sea, where they can dive up to a mile below the surface, holding their breath for a couple of hours. Once again, there are far fewer now than in the past, before sealers killed more than a million of them for their fat, which was rendered into oil; but fortunately they are still to be found on islands in many parts of the world.

Around Cape Horn in the South Pacific, over two thirds of the one hundred and fifty native species of plants on the Juan Fernández Islands are found nowhere else; and the islands are a seasonal home to shorebirds like the oystercatcher and land birds such as the paradise flycatcher, and home all year to the rare Juan Fernández hummingbird, found only on Más a Tierra. Seabirds are plentiful, as they are on most ocean islands, and the Juan Fernández petrel is indigenous there. It was probably a sailor in the southern seas who first named these birds after St. Peter, for when the sea is calm they look as though they are walking across the water, just as he was supposed to have done; and since they are often seen before a storm, they are sometimes called storm petrels. They have another name, a reminder of how intimately related the secular and the sacred were for those early sailors: Mother Carey's chickens, a rendering of *Mater Cara*, a name used by Spanish and Portuguese Catholics for Mary, mother of Jesus. And riding the currents around the archipelago are leatherback and great green turtles, but unlike the Dutch explorer Jakob Le Maire, they are able to make shore whenever and wherever they want.

The word "spice" is derived from the Latin *species*, and of all island species, spices have a special place in the human history of islands, island travel, and island conflict. Many of the most valuable spices, such as nutmeg and cinnamon and cloves and Jamaican allspice, are found almost exclusively on islands, and they sponsored island voyages for centuries, with the seafaring spice route rivaling the overland silk road. But since nothing lasts forever, the trade of spices was eventually surpassed by the market for fish. Cod fish. And nowhere in the world were there more cod than in the waters around Newfoundland.

Chapter Five

Amazing Islands

Real, Imagined, and In Between

"Newfoundland, an island on the east coast of America, opposite the gulf of St. Lawrence, is separated from the continent on the north by the strait of Bellisle, about twenty-one miles wide. It is of a triangular form [. . .] The length is about 380 miles, the breadth from 40 to 287, the circuit 900; and the surface is estimated at 35,500 square miles.

The shore of this island is lofty, desolate and rocky; but its serrated structure renders it abundant in excellent ports. The whole circumference of Newfoundland is indented, at intervals of two or three miles, by deep bays, having generally a smooth bottom, and a rivulet of pure fresh water at the head. It is in such situations the inhabitants have built their hamlets; their fishing craft being secured among the rocks [. . .] In regard to the hands it employs, Newfoundland is by far the greatest fishing station in the world."

—*Edinburgh Encyclopaedia* (1830)

"THE COUNTRY [NEWFOUNDLAND] yieldeth many good trees of fruit, as filberts in some places, but in all places cherry trees, and a kind of pear tree meet to graft on. As for roses, they are as common as brambles here: strawberries, dewberries, and raspberries [. . .] The timber is mostly fir, yet plenty of pineapple [pine] trees. There also be oaks and thorns [hawthorn]; there is in all the country plenty of birch and alder, which be the meetest wood for cold, and also willow, which will serve for many other purposes [. . .] I am informed that there are about 100 sail of Spaniards that come to take cod [. . .] besides 20 or 30 more that come from Biscay to kill whale for trane [oil . . .] and Portugals from Aviero and Viana and 2 or 3 ports more [. . .] The Englishmen, who commonly are lords of the harbors where they fish, do use all strangers' help in fishing if need require, according to an old custom of the country. As touching their tonnage, I think it may be near five or six thousand tonne [. . .] As touching the kinds of fish beside cod, there are herrings, salmons, thornback, place, or rather we should call them flounders, dog fish and another most excellent of taste called of us a cat [wolffish], oysters and mussels, in which I have found pearls [. . .] There are also other kinds of shellfish, as limpets, cockles, wilkes [snails], lobsters and crabs [. . .] There be fishes which (when I please to be merry with my old companions) I say do come on shore when I command them in the name of the 5 ports, and conjure them, by such like words. There also be the fishes which I may sweep with brooms on a heap, and never wet my foot [. . .] There be also other fishes which I tell those that are desirous of strange news, that I take as fast as one would gather up stones, and them I take with a long pole and hook [. . .] and with that tool I may take up in less than half a day lobsters sufficient to find three hundred men for a days meat [. . .] There are sea gulls, murres, ducks, wild geese and many other kinds of birds, especially at one island named Penguin."

—Letter from Anthony Parkhurst to Richard Hakluyt (1578)

Newfoundland, one of the twenty largest islands in the world, is sometimes called the Galápagos of geology, telling the story of the origin of the earth just as the story of the origin of species is told on those isles. Like Jamaica, its residents occasionally call it the Rock, scoured by glaciers into a tundra landscape of mosses and ponds and what one chronicler called its "mad indentations of harbour, cove and bay." Embedded everywhere, and in remarkable variety, are fossils (many from over 500 million years ago) that tell the story of a rich array of plants and animals that once lived on Newfoundland. And in the waters round there are gifts of grace, known for centuries to fishers from the Aboriginal Americas and from the coasts of Europe and Africa, who brought back catches so bountiful they almost swamped their boats. At the end of the fifteenth century, European traders and explorers trying to find a western route to bring spices from Asia heard of Columbus's failure to find a southern way, and so they looked instead for the northwest passage that had already found its way into seafaring fables. But not unlike Columbus, they found their way blocked by an island.

John Cabot, sailing in the service of Great Britain, reached this new-found island in 1497. Throughout the sixteenth century, it was called Terre de Baccalos, the land of the cod, as Britain competed with the Basque and the French and the Portuguese for control of the rich fisheries, which had been exploited by fishermen from the other side of the Atlantic since well before Columbus. But nobody had broadcast this fact, for like hunters and fishers everywhere, they wanted to keep the best locations (on what came to be known as the Grand Banks) to themselves.

The European fishing boats brought back harvests of cod that were both a gold mine to their sponsors (on the English boats the owners received two thirds of the catch, while the rest went to the crew; the Spanish and Portuguese paid wages) and a godsend to the majority of people in Europe, who did not have fresh meat to get them through the winter. Also, in Catholic countries, where fasting was prescribed every Friday (traditionally the day of Christ's crucifixion) as well as throughout the forty days of Lent and on other holy days—all of them days on which meat was forbidden—salt cod took on an importance that was almost religious.

As cod became profitable—even more profitable than the spices and sugar of the East and West Indies—the oldest continuous non-Aboriginal settlement in the Americas took shape on Newfoundland beginning in the sixteenth century. The men and women who settled there were first of all West Country English (along with French settlers) and then, beginning in the eighteenth century, many came from southeast Ireland; by the middle of the nineteenth century, over half the population was of Irish descent. They built "tilts"—rough huts of skins or boughs for seasonal shelter—and platform "stages" and "fish flakes" to clean, dry, and salt fish, along headlands with names like Bonavista and Blow Me Down. They navigated the fogs and furies of the North Atlantic in their flat-bottomed dories, rowing or sailing home from the "fishing grounds" (as they were called by those who brought the language of agriculture with them) through treacherous narrow straits between the rocky shores that guarded the harbors. They called these narrows "tickles," and they knew that the tickles were by no means the worst thing the sea had to offer.

It was to these same rocky shores that Scandinavians had come some five hundred years before Cabot, taking up residence on the northern tip of Newfoundland (at a place called L'Anse aux Meadows, which is now a UNESCO World Heritage Site as the earliest known European settlement in the Americas). The island, and indeed this very site, had also been visited a couple of centuries before that by the Dorset people, who hunted and fished down the North American coast from their home on the Arctic islands. In fact, a succession of Aboriginal peoples had occupied Newfoundland for millennia, from the early Paleo-Eskimo all the way to the Beothuk (of Algonquian heritage), who were living on Newfoundland when John Cabot sailed by.

The story of Newfoundland is of course much more than the story of cod. It is the story of islands, of land appearing out of the water and water overwhelming the land. It is the story of rocks shaped by the wind and the waves. It is the story of the various peoples who arrived by boat to make their island home on those rocks, settlers for whom seafaring was a way of life, and death. "Show me a map and I'll name you a dead man for / every cove between home and Battle Harbour," writes Michael Crummey in a poem called "'The Price of Fish' (September, 1887)." "I am twenty four years old, / there is no guarantee I will ever see twenty five." Or, in words that Walter Scott put in the mouth of a fishmonger speaking to a customer haggling over the price: "It's no fish ye're buying. It's men's lives" (in his 1816 novel *The Antiquary*). "We take our exuberance from what we have endured," says a contemporary Newfoundlander, which may be why the name of one of the most formidable island cliffs in the Atlantic, on the most easterly point in North America—Cape Spear on Newfoundland—was originally called *cabo da esperanca* in Portuguese, and then *cap d'espoir* in French. Another cape of good hope.

For centuries, probably millennia, Newfoundland has had a tradition of seafaring in the waters around its shores. And of storytellers like E. J. Pratt, describing "The Fog," which "stole in on us like a foot-pad, / Somewhere out of the sea and air, / Heavy with rifling Polaris / And the Seven Stars. / It left our eyes untouched, / But took our sight, / And then, / Silently, / It drew the song from our throats," while it "tangled up / The four threads of the compass, / And fouled the snarl around our dory"—the boats from which Newfoundlanders have fished for a long time, from close to home and also offshore on the open ocean. Good navigation, says the east coast writer Rob Finley, "relies on having good arguments for where you think you are, but a willingness to change your mind quickly when things take a sudden and surprising turn [and] you are severed from a world you had understood to be out there. All of the usual reference points become uncertain. You imagined a bell-buoy, the opening of a known harbour, a white house on a hill. Instead, a line of surf, a broken wall of cliff looming above you, or worst of all, sudden green water [signaling not an underwater atoll, but an iceberg]." That was, and still is, what seafaring is like around Newfoundland. An eighteenth-century rhyme, often referred to as "Wadham's Song" but probably memorized as a poem, would give the doryman directions. "From Bonavista Cape to the Cabot Isles / The course is north full forty miles, / When you must steer away North East / Till Cape Freels, Gull Island bears West North West. // Then North, North West thirty-three miles, / Three leagues off shore lies Wadhams Isles; / Where of a rock you must beware, / Two miles Sou'South East from off Isle bears. // The North West by West twelve miles or more, / There lies Round Head on Fogo Shore; / But Nor'Nor West, seven or eight miles / Lies a sunken rock near Barrack Isles. // Therefore, my friend, I would you advise, / Since all these rocks in anger lies; / That you may never amongst them fall, / But keep your luff and weather them all."

Wherever islands are, rocks (and reefs and sandbanks) big and small wait for the unwary, as seafarers have known since time began. And so have those who stayed at home, imagining both fatal encounters and favored isles like Thomas More's island Utopia, whose name means both "no place" and "a good place." The quintessential map of islands is one that recognizes that they have always been both real and imagined, the objects of both desire and dread. Oscar Wilde remarked that a map of the world that does not include utopia is not worth even glancing at. "It is not down in any map," says Ishmael, in Melville's *Moby Dick*, about the strange harpooner Queequeg's island home of Kokovoko. "True places never are." Newfoundland was a true place, long before it was on any map. It was a place to which some came looking for a home, and others for a harvest. Some came for spices, which turned out to be a world away. Others came to catch fish, and their middlemen made a fortune. Still others sought the fortunate isle of their dreams. At first sight, Newfoundland—with its terrifying cliffs, rocky shores, and scoured-out landscape surrounded by dangerous seas and a fearsome climate—wouldn't have seemed that sort of island, despite the rich bounty of the sea around it. Writing of its fisheries in 1801, Coleridge praised the island as "our best nursery for seamen. The severity of the climate, the perpetual fogs, the constant hardships, dangers and fatigues, render it a most laborious service. A sailor coming out of a Newfoundland fishing vessel and entering on board a man-of-war [warship] experiences an amelioration of his fate." There was even an island off the northern coast of Newfoundland called the Isle of Demons, where it was said there were not only bears and walruses but also gryphons, with the body of a lion and the head of an eagle, and other, even more malicious, presences. In reality, the frightening reputation of this demonic island, like that of many others before and since, was almost certainly the product of seafarers' imaginations, responding to the whistling and trumpeting and screaming of countless birds—gannets and guillemots and puffins and kittiwakes and auks (which were then still plentiful there). The birds would have greeted those who sailed by in the spring, when the ice pack was breaking up and the birds were breeding and nesting in their annual island round. Other strange sounds would have conspired as well, with the wind whipping the waves against the hollowed rocks to produce small explosions that sounded like half-heard cries from sea monsters. And there would have been sea lions honking and growling. The sailors would often hear this cacophony through the fog that was constantly closing in around them; their fears and fancies would do the rest. The imagination populated islands all over the world with strange and sinister spirits, and wove wonderful spells around others.

And yet, Newfoundlanders have always taken their prospects from the island they call home, and from the names they give it. "Outport is the characteristic Newfoundland word," the Newfoundlander Rex Murphy suggests (referring to the fjord-like fishing harbors that make up the island's coast), "the place removed and away, half sanctuary and half exile, sea-dependent and sea-conditioned, generous with challenge, scant with opportunity, a place at once of intense welcome and hospitality and yet desperately exposed, precarious and vulnerable. Newfoundland doesn't have outports. Newfoundland is the outport." An island of islanded communities.

But there have been other imaginings. In the early seventeenth century, George Calvert was granted a royal charter to establish a colony in Newfoundland as a refuge for Catholics facing persecution in England, and he called it Avalon, a blessed place, conjuring up the Garden of Eden and the Avalon of Arthurian legend—which was identified by many as Glastonbury in Somerset, with an island-like setting almost completely surrounded by marsh (and paradisaically named the "isle of apples" by the medieval chronicler Geoffrey of Monmouth). There, as legend has it, King Arthur had been taken to die—or to recover and return to conquer his enemies—and Glastonbury has for a long time been associated with the flowering of Christianity in Britain. Alfred, Lord Tennyson called it "the island valley of Avilion," where "falls not hail, or rain, or any snow, / Nor ever wind blows loudly; but it lies / Deep-meadow'd, happy, fair with orchard lawns / And bowery hollows crown'd with summer sea." It is the sort of place we all dream about, whatever our beliefs—a place of healing, in one version of Arthur's story, and a place of peace in another.

In 1594, Walter Raleigh said that should any harm happen to the Newfoundland fishing fleet, it would be the greatest calamity that could befall England. "The fisheries of Newfoundland are inexhaustible and are of more value to the empire than all the silver mines of Peru," added Francis Bacon, who had been given a substantial grant of land on the island in 1610. (Humphrey Gilbert had claimed Newfoundland for the British in 1583.)

The French had established an imperial presence on the continent of North America in the 1530s, and fished in the waters around Newfoundland throughout the next couple of centuries, curing and drying fish on the island. They also established small settlements there, with rights to fish along the northeast coast (which became known as the French Shore) confirmed by treaties with the British at the beginning of the eighteenth century.

Cod had been harvested in the North Atlantic (especially around Iceland) throughout the Middle Ages, and the waters around Newfoundland were legendary—which is to say their description, like that in many fishing stories, was not always accurate but invariably appealing. Yet even the most exaggerated tales weren't far from the truth. The shoals off the island—where the cool Labrador Current meets the warmer Gulf Stream, where the sea depths are manageable (since cod are by preference bottom feeders), and where food for the omnivorous cod was plentiful—became the greatest fishing ground the world had ever known. And the cod stocks seemed to be indestructible.

A series of fifteenth-century European maps had identified various islands west of Norway as the "codfish" or "stockfish" isles, referring generally to any whitefish, which are not as oily as other species and can be dried by cold air, sun, and wind on racks—the traditional method of processing fish and meat in northern regions, which did not have ready access to salt for preservation (unlike Portugal, for example) until the seventeenth century. After salt became cheaper and more easily available, fresh-served cod was almost a contradiction in terms. Dried salt cod, on the other hand—better known simply as saltfish—made an excellent meal. Once dried, cod has a protein content approaching 80 percent, and for centuries it fed much of the Western world, including the communities of enslaved Africans and transplanted Europeans on the islands of the Caribbean.

By the end of the sixteenth century, at least three hundred and fifty fishing boats were bringing back yearly harvests of cod to Europe, and from then until the late twentieth century, cod fishing both shaped and subsidized life on Newfoundland. The fishing took place both offshore, in larger ships and by fishermen called "bankers" (after the banks or shoals where they caught the cod), and inshore, closer to the island. The fishermen found their way by dory compass and dead reckoning, with the fog a constant menace (in Newfoundland speech, "it's mauzy almost half the year"). Ice floes, too, were a frequent hazard, alongside the ever-present dangers of wind and wave in boats filled to the gunwales with fish.

The habits of the local fishing communities in the outports, and the cod that sustained it, quickly became both the custom and the currency of the island; and fish were even used to pay school fees and taxes in the early eighteenth century. The bankers went to the fishing grounds in schooners, bringing along small dories that would then be launched around the mother ship in a starfish pattern to cover a wide area. The Europeans took their fish home "wet" or "green," to be salted and dried there; while the Newfoundland bankers and the inshore fishers (using longlines, or later traps and gill nets) would bring their catches back to the outports to be salted and dried and stored. Both inshore or offshore it was dangerous work, especially in the days of sail, and it required considerable skill; but it was also profitable, and created a unique island culture.

Cod was never the only seafood for Newfoundlanders, though it was certainly their inheritance. Inland there were salmon, various kind of trout, as well as eels and sticklebacks, while the ocean had herring and capelin, haddock, pollock (also known as Boston bluefish), mackerel, redfish (ocean perch), sturgeon, swordfish, tuna, halibut, and turbot (sometimes called Greenland halibut), along with crab, shrimp, scallops, and lobster. Whales—blue and right, humpback and minke—were hunted from the sixteenth to the twentieth centuries. And seals, of course, which continue to be taken on the pack ice.

But cod was *the* fish, and both the economy and the identity of Newfoundland depended on it. And it wasn't only cod meat that was prized. Cod-liver oil became a valued commodity, and Newfoundland's contribution to the 1851 Great Exhibition of the Works of Industry of All Nations (taking place in London, in the famed Crystal Palace) was a bottle of cod-liver oil. (Iceland provided a sample of wool; Jamaica displayed artificial flowers made from yucca fiber, Jamaican plum and pimento [allspice], and a walking stick made in London from the pimento tree; while Tahiti offered a translation of the Bible into the Tahitian language. The Galápagos had no display, but then again, Darwin's *On the Origin of Species* was not published until eight years later.)

But something about cod-liver oil should have alerted those watching the ups and downs of the codfish catches. Beginning in the nineteenth century, fishermen had noticed that the oil content of cod livers depended on where the fish were caught. This worried them, and they wanted more attention paid to local habitat and spawning cycles, to counter the widespread belief that cod roamed throughout the northern Atlantic relatively unaffected by regional catches. Even as notable a scientist as Darwin's friend Thomas Huxley proclaimed (in his address to the 1883 Great International Fisheries Exhibition, also in London) that, while it might be possible to exhaust some inland fish species such as river salmon, the number of ocean fishes such as cod was "so inconceivably great that the number we catch is relatively insignificant, and [. . .] the magnitude of the destructive forces at work upon them [in their natural environment] is so prodigious that the destruction effected by the fishermen cannot sensibly increase the death rate."

Huxley's prognosis was based on the simple but sadly mistaken premise that fishermen were individuals who on their own could not possibly inflict serious injury on the large and widely dispersed cod community. This account, though picturesque, ignored the fact that cod are, more or less, a local fish, like herring (or, to take a land species, caribou), and cod families flourish in specific places under particular conditions. If overfished in one location, or if their food supply there is damaged, numbers may not recover. The debate continued into the twentieth century, modified in the 1930s and 1940s by a slightly more sophisticated notion of regional cod stocks, but still obscured by cycles of migration from one area to another. To complicate the debate, the use of trawls and nets inevitably resulted in bycatches, where fish other than those for which the fisher was licensed were caught—and then thrown back dead to avoid fines. The situation became even more damaging after the introduction of synthetic instead of natural fibers for the nets. These were stronger, allowing for bigger nets; but since they are not degradable, if they were lost during a storm they would drift forever, ghost nets in the sea, catching whatever came their way.

But years earlier, in the 1920s, the cod's fate had already been sealed, with the invention (by the American Clarence Birdseye) of a method of quick-freezing fish. That enabled the use of large so-called factory ships, with massive open-stern equipment for trawling on the Grand Banks, resulting in catches previously unheard of. Many of Newfoundland's inshore fishers, who could see the end of the line, called for restrictions, while conflicts between urban and outport interests on the island also contributed to the devastating decline of fish stocks. And then there was another element. Just as residents of large islands often discredit those of small islands, so mainlanders often discount *all* islanders; and when Newfoundland became part of Canada in 1949, a set of national and scientific preoccupations from the mainland were put in play on the island. The end was near, for this thick weather of reasons—and the commercial cod fishery in Newfoundland was closed in 1992.

This brought back one of the oldest and most familiar questions in the history of island settlement: to leave or to stay? Leavers and stayers have peopled Newfoundland stories for centuries, and a song called "Sonny's Dream," by Ron Hynes, has become an island anthem, with versions of it sung in Ireland as well. "Sonny don't go away," his mother cries. "I am here all alone / And your daddy's a sailor who never comes home [. . .] Sonny carries a load though he's barely a man / There ain't all that to do, still he does what he can / And he watches the sea from a room by the stairs / And the waves keep on rollin', they've done that for years [. . .] He's hungry inside for the wide world outside / And I know I can't hold him though I've tried and I've tried."

The sailor was standing on a mountain, on an island far out in the ocean. He was marooned, and he was not very happy about it. So he wrote a poem. "I am monarch of all I survey," it begins. "My right there is none to dispute; / From the centre all round to the sea, / I am lord of the fowl and the brute. // Oh, solitude! where are the charms / That sages have seen in thy face? / Better dwell in the midst of alarms, / Than reign in this horrible place. // I am out of humanity's reach, / I must finish my journey alone, / Never hear the sweet music of speech; / I start at the sound of my own. [. . .] // Society, friendship, and love, / Divinely bestow'd upon man, / Oh, had I the wings of a dove, / How soon would I taste you again!"

The sailor's name was Alexander Selkirk, and his story was the inspiration for Defoe's novel *Robinson Crusoe*. However, the author of this poem, titled "Verses supposed to be written by Alexander Selkirk," was not Selkirk but the English poet William Cowper, and the "supposed" was just a literary convention—like Defoe's claim that his novel was written by Crusoe. The popularity of Cowper's island poem went far beyond literary circles during his lifetime. In fact, in its day these verses were almost as famous as the song "Amazing Grace" is in ours.

And there was a connection. Cowper's name may have faded, but in his lifetime—the latter half of the eighteenth century—he was very well known. No less a literary critic than Coleridge called him "the best modern poet"; and Cowper had become a household name because of his collaboration with the Reverend John Newton in writing a series of hymns to go along with weekly sermons in the little village of Olney in England where both men lived. One of them was Cowper's celebrated hymn "God moves in a mysterious way, / His wonders to perform; / He plants His footsteps in the sea, / And rides upon the storm." Another, by John Newton (a former sailor and slaver), was "Amazing Grace," and it eventually set the all-time record for crossover hits; but in its time Cowper's 1782 poem about Selkirk topped the charts, so to speak (even if, perhaps, it wasn't as fine as another island poem of Cowper's titled "The Castaway"). It was the sound of a voice, an island voice, crying in a watery wilderness where fate conspired against free will—and where home was poised uneasily between dream and nightmare. Cowper himself was having religious doubts, and suffering from severe depression, and so he was all too familiar with feeling like a virtual castaway; and Selkirk, the actual castaway, had said that the Bible saw him through his hardship. Island isolation—"half sanctuary and half exile," in Rex Murphy's words.

The word "maroon," associated with an enforced island existence, comes from the Spanish *cimarron*, meaning wild; and *Robinson Crusoe* (its full title begins *The Life and Strange Surprizing Adventures of*

Robinson Crusoe, of York, Mariner) was about freedom in wildness as much as it was about island confinement. But there was something else, an essential island contradiction, for islands may offer not only solitude but company. Close company. And Crusoe had a companion, as well as unwelcome visitors. Along with lives of adventure and survival, the eighteenth- and nineteenth-century European literature of islands explored the conditions of alienation and affection, the savage and the civilized, the individual and society, with them all changing places like dancers in a quadrille—at any given moment you couldn't be sure what would be partnered with what. Contemporaries recognized this, and perhaps because many of them knew about islands, they saw more to Selkirk's and Crusoe's island story than a "boy's own" adventure. When Richard Dana passed by the Juan Fernández Islands on his way from Boston to California, he said that Más a Tierra—now called Robinson Crusoe Island—had "the sacredness of an early home" for the sailors who regularly stopped there. (He wrote about this voyage in 1840 in his best-selling book *Two Years Before the Mast*; "before the mast" referred to the front of the ship where the common sailors slept, a social island on board.)

Alexander Selkirk was a Scot who, like many of us at one time or another, didn't get along with family and friends. After being called to account for bad behavior in church—a serious matter in many communities in seventeenth-century Scotland—he picked up and signed on a ship, sailing with a privateer over several of the seven seas. Old habits die hard, and Selkirk had as tough a time getting along with his shipmates as he had with his family; but (also like many of us) he dreamed about going out on his own, being free and independent, able to do whatever he wanted without worrying about others, domesticating the world according to his own wants and needs. He began feuding with the crew and the captain, and when they stopped in the Juan Fernández archipelago in the fall of 1704 for a few days rest and repair, he said he wanted to stay there awhile and wait for the next ship to come along. As the ship weighed anchor, Selkirk had second thoughts, running down to the shore and calling out to the crew to wait for him. Unfortunately, they didn't like him any more than he liked them, and their laughter rang in his ears for four years—until, finally, another ship came by and picked him up.

Although the Pacific is a very big ocean, familiar routes for European sailors made it seem much smaller back then, with customary courses becoming what one mariner referred to as well-traveled turnpikes. William Dampier was the navigator of the ship that eventually picked Selkirk up (and had been with the fleet four years earlier when Selkirk went ashore); he was quite a good writer but, as mentioned, it was the account of Selkirk's ordeal by Woodes Rogers, the captain of the ship that found the marooned Scotsman, that inspired Defoe. For his part, Defoe blended fact and fiction in a manner that all but defined the novel as a literary form, transforming Selkirk's temperate Pacific island into a tropical Atlantic setting with just enough carefully staged challenges to make the natural conditions of survival and sovereignty coincide with the carefully nurtured colonial and commercial enthusiasms of the age. *Robinson Crusoe* was one of the earliest novels in the English language, and is still one of the most widely read, with our contemporary reality TV shows and survivor sagas being popular imitations. And it is no small tribute both to Defoe's genius and to the appeal of island chronicles that the story has been taken up in our supposedly postcolonial and postindustrial era by writers such as the Nobel laureate Derek Walcott (who, when he grew up on St. Lucia, had a picture of Robinson Crusoe on his bedroom wall) in a book called *The Castaway* (1965), which includes a long poem titled "Crusoe's Journal," and in a play about Crusoe and Friday called *Pantomime* (1978); in a novel called *Foe* (1986), a play on Daniel Defoe's name, by another Nobel Prize winner, the South African J. M. Coetzee; and in *Vendredi ou les Limbes du Pacifique* (1967; the English title is *Friday*), a novel by the French writer Michel Tournier.

Robinson Crusoe was a phenomenon in its time—and it still is. Coleridge said that he read *Robinson Crusoe* before the age of six; and later in the nineteenth century a survey of children in Great Britain found that *Robinson Crusoe* was one of their two favorite books. The other was *Swiss Family Robinson* (1812), which was intended to teach social responsibility and self-reliance and reflected some of the cooperative idealism of *Robinson Crusoe*—though *Swiss Family Robinson* turned the tale into a manifesto of family values. (The original German title, *Der Schweizerische Robinson*, translates as "Swiss Robinson," which is to say a Swiss version of *Robinson Crusoe* rather than a Swiss family named Robinson.) First drafted by the Swiss pastor Johann David Wyss, it was completed by his son Johann Rudolf Wyss and then abridged in the English translation, eventually going through so many adaptations that the original is mostly obscured (though William Godwin did a reasonably faithful translation a few years after the book was first published). It was enormously popular as a school text in the late nineteenth century, along with Frederick

Marryat's *Masterman Ready* (1841), said to have been written at the request of his children as a sort of sequel to *Swiss Family Robinson.* Further island tales included Alexandre Dumas's *The Count of Monte Cristo* (1844), with an island setting near Corsica and Elba; Robert Michael Ballantyne's *The Coral Island: A Tale of the Pacific Ocean* (1858); Jules Verne's *The Mysterious Island* (1874); and Robert Louis Stevenson's *Treasure Island* (1883). And it was not just young readers who were fascinated by these works. The first books that Oscar Wilde asked for after his first trial, while awaiting bail in London's Holloway Prison, were by Stevenson.

Is separateness, or togetherness, the condition of human existence? This was the question that had been posed by John Donne in the seventeenth century, and it was taken up in the eighteenth century by Defoe and the new novelists who wrote for an audience that embraced a class-based individualism in which the sense of a fundamentally inhospitable world, requiring human ingenuity to make it habitable and sociable, was taking hold. It was not a new notion—that was the story of Noah as well, after all—but it seemed to have the philosophers onboard, and it resonated with the new interest in individual identity and social relationships. Capitalism was picking up steam as a political and moral ideology, and it would be given economic credibility later in the century by Adam Smith in *The Wealth of Nations* (1776). Defoe's novel became the case book and the poster child for entrepreneurial initiative, coinciding with the rise of what came to be called the middle class.

But utopian socialism was also popular in philosophical circles in the late eighteenth and early nineteenth centuries, having had a run in the sixteenth and seventeenth centuries with Thomas More's

Utopia (1516) and Francis Bacon's *New Atlantis* (1627), both set on islands (Bacon called his Bensalem). In his famous tract *Emile* (1762), Jean-Jacques Rousseau proposed that *Robinson Crusoe* "supplies the best treatise on education according to nature. This is the first book Emile will read; for a long time it will form his whole library, and it will always retain an honoured place. It will be the text to which all our talks about natural science are but the commentary. It will serve to test our progress towards a right judgement, and it will always be read with delight, so long as our taste is unspoilt." Capitalists and socialists both claimed the island as the archetype—an oasis of competitive enterprise in a desert of uncivilized instinct, or a haven of natural cooperation in a sea of debilitating struggle.

Hovering over all these accounts, and defying classification, was Shakespeare's play *The Tempest*, with the magisterial Prospero as what one modern writer called "the great overseer of the imagination in Western literature" trying to shape island reality—in the form of Caliban—in his own image and in his own language. The novelist John Fowles described Shakespeare's play as "the first guidebook anyone should take who is to be an islander; or since we are all islanders of a kind, perhaps the first guidebook"; and the literary critic Northrop Frye began *The Educated Imagination* (1963), his classic account of the relationship between literature and life, by asking us to suppose that we are shipwrecked on a South Sea island. "The first thing you do," he suggested, "is to take a long look at the world around you, a world of sky and sea and earth and stars and trees and hills. You see this world as objective, as something set over against you and not yourself or related to you in any way [. . .] So you soon realize that there's a difference between the world you're living in and the world you want to live in. The world you want to live in is [. . .] not an environment but a home; it's not the world you see but the world you build out of what you see. You go to work to build a shelter or plant a garden [. . .] transform[ing] the island into something with a human shape." That's the story of Robinson Crusoe. It is the story of what it is to be a thinking and feeling person in the world. It is the story of human beings and islands.

Stories of sailors held hostage by sensuous island seductions and strange islander spells are part of a myth of islands that has become famous in the world's literature, as well as in its history. The contradictions of island habitation include the perverse scientific curiosity on the *Island of Dr. Moreau* (H. G. Wells's 1896 science-fiction novel), the scary sociopathology of William Golding's *Lord of the Flies* (1954), the season spent on the Isle of Skye in Virginia Woolf's *To the Lighthouse* (1927), and the artistic inspiration found by the writer John Millington Synge around 1900 in the hardscrabble life—and death—of the folk on the Aran Islands off the west coast of Ireland.

Crime and mystery writers have had a fondness for islands, with Agatha Christie setting a number of her stories there. *Evil Under the Sun* (1941), for example, takes place on a small tidal island called Burgh on the coast of Devon and features the fastidiously eccentric detective Hercule Poirot. His fellow Belgian, the writer Georges Simenon, sends his Parisian inspector Jules Maigret to the Mediterranean island of Porquerolles (in *Mon ami Maigret*, 1949), while Stieg Larsson's *The Girl with the Dragon Tattoo* (2005) begins on the fictional Swedish island of Hedeby. Other mystery novels throughout the ages have centered the action in isolated or "islanded" locations, including the castles, convents, and monasteries of Gothic fiction.

Agatha Christie thought for awhile about buying an island that she could sail off to every now and then when she needed to be alone. But the islands nearby were rocky, with only trees and birds and maybe a few sheep. So she put aside the idea. "No," she said. "An island is, and should be, a dream island [. . . with] white sand and blue sea—and a fairy house, perhaps, built between sunrise and sunset: the apple tree, the singing and the gold." Avalon, that is—a temperate garden of Eden with an apple tree, rather than a tropical paradise with a palm tree.

Some of the most popular children's books in Western literature have also been set on islands, or have islands as the center of the story. J. M. Barrie, at the beginning of the twentieth century, has Peter Pan living with his gang of Lost Boys and the fairy Tinker Bell on the island of Neverland, and Hugh Lofting's *Doctor Dolittle* series (which began in 1920) includes several island adventures. Lucy Maud Montgomery set her celebrated tale of the orphan *Anne of Green Gables* (1908) on her native Prince Edward Island, transforming it into an idyll of innocent adolescence. In Astrid Lindgren's widely adapted series of mid-twentieth-century books featuring Pippi Longstocking, Pippi's father is a sea captain who is washed ashore on an island in the South Pacific known as Kurrekurredutt Isle, where he is in danger of being eaten by the natives, but instead they make him king. Maurice Sendak's classic *Where the Wild Things Are* (1963) places the Wild Things on an island. *Call It Courage* (titled *The Boy Who Was Afraid* in England), written by Armstrong Sperry in 1941 and since translated into over twenty languages, is set on Hikueru, one of the Tuamotu atolls in the Pacific Ocean.

The popular British writer Enid Blyton began several of her series of children's books with island adventures—*The Secret Island* (1938), *Five on a Treasure Island* (1942), and *The Island of Adventure* (1944)—while the idea of an "island" of Gaulish resistance surrounded by Roman army forts is at the heart of the comic-book series *Asterix*, written by René Goscinny and illustrated by Albert Uderzo. Two real islands visited by Asterix and his friend Obelix over the years are Corsica and Britain. The Belgian comic-book artist Hergé (Georges Remi), too, features islands in his series *The Adventures of Tintin*, including a mysterious Scottish island (*The Black Island*, 1938) and a volcanic island in the Java Sea (*Flight 714*, 1968).

Many other books and films appealing to young and old alike have islands as their setting, from *King Kong* (notable citizen of Skull Island in the Pacific) to the genetically cloned dinosaurs in *Jurassic Park*, located on a fictional island off the coast of Costa Rica. The action—and often the evil enemy—in Ian Fleming's James Bond books and films almost always involves an island, and indeed Fleming himself retreated from one island home (England) to another (Jamaica) to write his books, in a place called Goldeneye (which is now owned by Chris Blackwell, whose Island Records has produced notable island musicians such as Bob Marley and U2). From its early days, radio and television took up the island theme with series like *Gilligan's Island*, *Fantasy Island*, and *Survivor*. And *Desert Island Discs*, the BBC radio show that began in 1942, in which famous people choose the few favorite things—a book, one "luxury" item, and some music—that they would take with them to a desert island, is one of the longest-running shows in radio history.

The Broadway musical *South Pacific*, which one writer calls the *Robinson Crusoe* of the postwar years, was first staged in 1949, with a very popular movie a decade later. It was based on two stories by James Michener, "Fo' Dolla'" and "Our Heroine," from his *Tales of the South Pacific*. While some say the stories had their origin on the island of Espiritu Santo, Michener himself spoke of seeing the name "Bali Ha'i" on a cardboard sign on Mono Island, and of being inspired by friends in Guadalcanal (both part of the Solomons). Others have noted the similarity of Bali Ha'i to the island of Nuku Hiva, on the Marquesas. Which is to say, it could be both anywhere and everywhere—or nowhere—in the South Pacific; but the search for its source is appropriate since, in the words of the musical, "Bali Ha'i mean I am your special island."

Special islands come in every shape and size, and some of them are man-made. In 1964, Ernest Hemingway's younger brother, Leicester, anchored a bamboo raft with an engine block on a shallow shelf in international waters off the coast of Jamaica and called it the island republic of New Atlantis. He sought diplomatic recognition for what he described as "a peaceful power" (with seven citizens) and a currency that he called the scruple. He promised that it would not threaten its Caribbean neighbors, and dedicated its resources to protecting fish stocks in the region. It had a limited shelf life, as utopias sometimes do, disappearing in a storm several years later. But in its own way it was a seagoing counterpart to the raft in the middle of the Mississippi River that was an island home to Huck and Jim in Mark Twain's *Adventures of Huckleberry Finn* (1884). An in-between island.

Sometimes an island is in between the comic and the tragic, as when pirates arrive, finding relatively safe haven on islands that symbolize their unlicensed or outlandish existence. Piracy probably began shortly after the first boat set out with valuable commodities (including important people) thousands of years ago. Mediterranean and North Atlantic islands were favorite hunting grounds for the Vikings and other European (as well as North African) pirates, and the Indian and Pacific oceans provided both safe havens and commercial strongholds for pirates, with piracy in the South China Sea flourishing from the 1500s to the 1800s. Islands became centers of storytelling about their exploits, along with those of legendary seafarers such as Sinbad the Sailor. Jamaica and the other Caribbean islands, with the invitation they offered to exploit divisions between British, French, Dutch, and Spanish authorities, were a gift to pirates from the sixteenth to the eighteenth centuries, some of them moving back and forth between piracy and privateering (which was often little more than piracy sponsored by governments). Smuggling, piracy's close cousin, was also often centered around islands, with Smuggler's Coves almost everywhere on the coastlines of the world. And while an island such as Newfoundland was never the haunt of pirates, a wry account of its history of political chicanery by Rex Murphy brings it into that domain and reminds us, too, that the social harmonies and noble democracies that evolved on some islands (such as Tahiti and Iceland) are only part of the island story. "Pirates visited other territories to conduct their proper business, to loot; they came to Newfoundland as to a finishing academy or a spa for the profession. More for study than rapine, they bent themselves to close watch of the Order of Newfoundland Politicians—the platonic original of freebooting and sleveenery."

It is no surprise that pirates found a treasured home in the imagination, just as islands did. Their place in storytelling probably goes back to its beginnings, where rascals and ruffians always made for a good plot. But there is something more at work—something that has to do with the elusive reality of islands and island life. From Stevenson's *Treasure Island* to science-fiction pirates capturing merchant spaceships, there is a stubborn storytelling fascination not only with buccaneers but also with their island retreats, the "probably–probably not" storyline of buried treasure given an historical turn with tales of the Scottish pirate-privateer Captain William Kidd in the seventeenth century and the French (or possibly Haitian) Jean Lafitte a century later. Adaptations of pirate stories such as Gilbert and Sullivan's nineteenth-century comic opera *The Pirates of Penzance* and the twenty-first-century *Pirates of the Caribbean* films also illustrate how easily such stories lend themselves to irony and satire, which may be another turn on the confusions of reality and the imagination that islands so often display.

Sometimes, the specialness—or strangeness—of an island gives it a nameless identity, like Utopia itself. There is a lake island on the Pacific coast of Canada, once the location of a whaling shrine for the Nuu-chah-nulth (or Nootka) Indians, that catches this quality. Whaling had become an important commercial enterprise in the nineteenth century, with indigenous people throughout the South Pacific taking to its long voyages and uncertain conditions with the craft and courage that had taken their ancestors to the ocean islands. But further north, the Nuu-chah-nulth had hunted grey whales in the spring and humpbacks in the summer for well over a thousand years, and they may be among the earliest whale hunters of whom we have a record in the oral tradition. The community depended on whaling not only for food and materials for shelter and clothing and tools, but also for trade and the social, economic, and political prestige of a valuable commodity. Though salt and spices may have been the ancient equivalent of modern oil and gas in the global market, for centuries whales and seals and sea otters—and of course cod—rivaled them.

Like whale hunting itself, the spirits that govern it are dangerous; and the rituals that were necessary to keep them appeased were both secret and separate from any other ceremonies in the community. For the Nuu-chah-nulth, they took place on a little island—an island on a lake, not in the ocean where the whales were—near the village of Yuquot. The lake was named after John Jewitt, a metalworker who specialized in crafting armor and (after he became fascinated with the stories of the sea that Captain Cook and others were bringing back) making hatchets and daggers to trade in coastal communities. He was captured when the ship he was traveling on, sailing out of Boston in 1803, was overpowered at anchor by the Nuu-chah-nulth. One of the few to survive, he became the European slave of the chief, Maquinna, and remained with him for two and a half years, until he orchestrated his release on the visit of another trading ship, subsequently writing one of North America's best-known captivity accounts, *A Narrative of the Adventures and Sufferings of John R. Jewitt* (1815). By his time, harpoons were the whale-hunting weapon of choice, and Jewitt's expertise in crafting metals made him the maker of the chief's harpoons, for hunting whales was hierarchical with the Nuu-chah-nulth, the sole prerogative of chiefs.

The ceremonial practices and protocols that made the lake island such an important place in Nuu-chah-nulth life are mostly lost, though we do have a few descriptions and some early photographs of the whaling shrine. There were wooden replicas of deceased hunters there, along with what was described (by an early-nineteenth-century sailor, almost certainly from hearsay) as a gallery of human bones marking the limits of the sacred space, and large carvings of whales, with human skulls arranged on the back of each. An "island of the dead" it was called in a film reconstruction by the great photographer Edward S. Curtis in 1915. A few years earlier, in one of the most infamous acts in the history of anthropology, the island shrine had been acquired by Franz Boas (through his local Tlingit/Kwakwaka'wakw intermediary) for the American Museum of Natural History in New York, an island of another kind. The shrine remains there, mostly in storage, despite Boas's ambition to have it displayed as the centerpiece of the museum's collection. Nuu-chah-nulth elders who have visited, as well as those who have never seen it, are divided about its return, for the spirits need to be properly treated, and those who knew the proper ways and means have long since passed away. The lake water around the island defined a symbolic distance, a spiritual rather than a physical boundary, for the island was only a short distance from land, and easily accessed; but it was a boundary marking a place of dangerous power.

Particular islands—Easter Island is a haunting example—remind us of both the permanence of religious belief and the transitory nature of life. On Loch Maree, a lake in the northwest Highlands of Scotland, there is an island called Isle Maree, which gives this a poignant human turn. It was said to be a place of healing; in the nineteenth century, at a time when there were no psychiatric hospitals, it was believed to offer a cure for "lunatics," as the mentally ill were then cruelly called. Sufferers were brought to the lake, put into a boat, rowed to the island, and then dropped into the water with a long rope tied around their waist. They were dragged as quickly as possible around the island three times in a clockwise direction (which was apparently an important part of the cure), and in the words of one of the old-timers, "if the patient was still alive and capable of swallowing anything he was landed on the island, and as if he had not got already more than sufficient water inside him he was made to swallow a lot more from Naomh Maolruaidh's (Saint Malrubas's) holy well" on Isle Maree.

Unholy islands, like holy ones, have always been out there; sometimes, almost as witness to their magical or mystical power, they change from one to the other. Robben Island, at one time a leper colony and then the prison where Nelson Mandela and his freedom fighters were incarcerated, has now become an emblem of liberation. John, the author of the biblical book of Revelation, was sent to the island of Patmos in the Aegean Sea as a prisoner to work in the local salt mines. A thousand years later, a monastery was founded there in his honor, and it became a place of pilgrimage.

But many islands were simply places to put away outlaws and outcasts. Leper colonies, with their exemplary threat of contagion, were routinely isolated on islands such as Culion in the western Philippines, Fantome Island off Australia's Queensland coast, and Chacachacare between Trinidad and Venezuela. And so were so-called lunatic asylums and prisons. Perhaps the most spectacularly distant was the island continent of Australia, used by Great Britain during the eighteenth century as a penal colony; and then—just to show that you can never have enough islands for the convicted—little Norfolk Island, a thousand miles east of Australia, was set up as a penal colony for the penal colony. Over the centuries, several notorious prison islands have been almost as far away from "civilization" as Australia was from Europe back in the 1700s. St. Helena is over eight hundred miles from the nearest land—which is in fact another island, Ascension—and twelve hundred miles from the mainland of Africa. It was the final prison exile of Napoleon after his defeat at Waterloo in 1815. He had escaped from his exile on the Mediterranean island of Elba—which is just a few miles from the coast of Italy—and his enemies were not about to make the same mistake twice. So Napoleon became perhaps the only emperor in modern times to be born on one island (Corsica) and die on another. The British later used St. Helena to imprison a Zulu king (from 1889 to 1898) and as a prisoner-of-war camp during the Boer War.

Napoleon holds a spell over yet another island, San Michele in Venice. The site of the first Renaissance church in the city, it became a prison in the eighteenth century; but when the region was occupied by Napoleon's troops in the early nineteenth century, an order was passed requiring Venetians to stop burying their dead in relatively shallow graves (because of water levels) right in the city. A cemetery was created on the outlying islands of San Cristoforo and San Michele (now joined together). Widely known as Cemetery Island, it is the final resting place of such notables as the dance impresario Sergei Diaghilev, the composer Igor Stravinsky, and the poets Joseph Brodsky and Ezra Pound.

There are islands whose reality defies even the grimmest imagination when it comes to isolation, deprivation, and wickedness. Three of them, initially called the Triangle Islands, were taken over by France after the Treaty of Paris in 1763 (along with Guadeloupe and Martinique). They are a few miles off the coast of French Guiana in South America, and became known as the Devil's Islands because of the currents that swirl around them. Their name was changed again to the Salvation Islands, to encourage settlement, but this was a spectacular misnomer—since a good part of the settlement was for a prison. This penal island could not shake off the devil's name and reputation, presumably because the pretense was just too grotesque. It was to this Île du Diable that the French army officer Alfred Dreyfus was sent in 1895 with a life sentence, after his wrongful conviction for treason—precipitating a political scandal that attracted international attention. Dreyfus languished on the island for four years, until his presidential pardon in 1899. One prisoner who followed him there told of spending his time talking to the sharks that surround the island; and it became famous again in a story (and movie) about Papillon (the Butterfly), a prisoner who claimed to have escaped. It's a good story, combining fact and fiction in the best tradition of island novels. The prison was closed down in the 1950s, and today Devil's Island is a tourist resort, its isolation and the dangerous waters around it now protecting leatherback turtles, one of the largest living reptiles in the world.

Fort Jefferson, on a coral outcrop in the Dry Tortugas (*tortugas* means tortoises in Spanish), at the very end of the Florida Keys, offers another model of an island stronghold. The fort was begun in the 1840s, to protect American shipping lanes between the Gulf of Mexico and the Atlantic Ocean. It became the largest brick fortress in the Americas, but it ended up being used as a military prison, and is now a national park and wildlife sanctuary. And there are plenty of islands in the Caribbean that feature the remains of a fortress, built during the centuries when the various European powers played musical chairs in the region. Many of these former military strongholds are public parks today.

Some prison islands were chosen not for their isolation but for their proximity, both in order to service the islands easily and as a cautionary example for mainland renegades. New York City's Rikers Island has been used as a prison for many decades, and Roosevelt Island (once called Blackwell's Island, and later Welfare Island), located in the East River between Manhattan and Queens, served as both prison and a mental hospital. One of the most infamous prison islands in the Americas was named by a Spanish sailor who was lost in a fog—and only saved from shipwreck by pelicans flying about and warning him of the rocks ahead. He named the island *La Isla de los Alcatraces*, the Island of the Pelicans, in gratitude: Alcatraz, the maximum-security prison in San Francisco Bay. Its nickname, yet again, is The Rock.

Along the coasts of Africa are the islands of the slave trade, such as Zanzibar in the Indian Ocean, which has been occupied for millennia by humans. In modern times, it became famous for its trade in spices and infamous for its slave market, which remained in business until the 1870s and through which as many as fifty thousand enslaved Africans passed every year. On the Atlantic side of the continent, the Portuguese Cape Verde Islands were exemplary in the worst way: the starvation and sickness that was endemic from the first settlement of enslaved Africans there in the fifteenth century was replaced by the brutality and bewildering excesses of a plantation system established on Cape Verde and then exported around the world. Cape Verde was hardly unique, but it was by all accounts a horrid place, grimly remembered as a staging point in the chronicles of slavery. And off the coast of Senegal is a small island called Gorée. It was one of the first places on the west coast of Africa to be settled by Europeans—first the Portuguese, then the Dutch, and then shuttling between them until the British and finally the French got into the game in the seventeenth century. It was not a major slave center, but its infamous House of Slaves, built in the 1780s, has become an island memorial to the holocaust that was the African slave trade.

Humans live by both their ambitions and their anxieties, their fancies and their fears. Islands are good places to send people you never want to see again—and they are also good places to go if you never want to see people again. Both Alexander Selkirk and Lemuel Gulliver yearned for their island "home" after they had spent some time back in society; and D. H. Lawrence's short story "The Man

Who Loved Islands" (1928) tells of a man who moves to smaller and smaller islands, seeking one island that will be "a nest which holds one egg, and one only. The islander himself." This man didn't even want trees or bushes on his island. "They stood up like people, too assertive." He ends up, not surprisingly, about as happy as Gulliver is at the end of his travels. In *A Small Place* (1988), Jamaica Kincaid does a good job of turning her native Caribbean island of Antigua into a place you want to get away from. She did—but like many islands, it still seems to have a hold on her.

Islands have often accommodated a wide range of human wants and needs. There have been (imaginary) islands where only women lived, conceiving children by looking at themselves in a mirror or being impregnated by the wind; and there were isles exclusive to men. Marco Polo mentioned both, describing how the men traveled from their island to the women's and stayed for three months of the year. He located them near Socotra, but others mapped them in the waters of Southeast Asia; and while some of these island accounts may have been as reliable as Gulliver's, Columbus and his seafaring contemporaries were inspired by a few of them—as indeed were sailors from Asia and Africa, who took to sea because of similar stories and songs. Some imagined islands even found their way onto maps; for instance, one of the islands of women, called Imangla, was featured on a map from 1575 at the western edge of the Pacific—at a time when the ocean was mostly empty on such maps, though filled with fanciful creatures and fabulous symbols—along with Inebila, inhabited only by men. Famous for its feminine ideals was the real island of Lesbos in the Aegean, which legend (as told by Diodorus of Sicily) also identified as an occasional home for the fierce race of women called Amazons. Another real island called Engano, south of Sumatra, was said by Malay sailors to be populated only by the fair sex—but no man dared venture there to verify this. Trobriand islanders told about a dangerous island of women, though it was far away. And among the Inuit there is a story about an island where women flee after killing the murderers of their husbands, but find themselves alone there with only their sons. When a man lands on the island unexpectedly, they smother him to death in their excitement.

The legendary Hy-Brasil was first marked on the charts in the fourteenth century as being located off the coast of Ireland—and remained there for over five hundred years. Landings were recorded, mainly by Irish monks and other seafarers, who brought back stories of a promised land, though there were other stories about Hy-Brasil being shrouded in fog, or perhaps even beneath the sea, and appearing only once every seven years. The island was a phantom, but like Atlantis a persistent and appealing one. Hy-Brasil bears no relation, real or imagined, to what became the country of Brazil, which was in fact first called the Island of the True Cross (and then later named for the wood *pau-brasil*, or brazilwood, found there) by the Portuguese explorer Pedro Álvares Cabral, who came by in 1500—on his way to the Spice Islands, of course.

And then there was the curious island of Buss. It was first reported in the North Atlantic in 1578 by a passenger on one of the ships—a sturdy type of boat used in the herring fishery, called a busse—that the English explorer Martin Frobisher had put together for his third voyage to find a northwest passage. Buss was entered on nautical charts, and then assigned to the Hudson's Bay Company to harvest furs. But it, too, did not exist. Perhaps it was a harbinger of certain business practices—or a reminder that maps and myths have more in common than we suppose.

There are circumstances when the "observation" of imagined islands has a real explanation, such as icebergs at a distance in the northern or southern oceans, or the white scum that is created when palolo worms rise to the surface in the Pacific, or masses of pumice and mats of vegetation that carry nomadic flora and fauna across the ocean in the mysterious story of island travel. But other sightings are harder to explain. Three Aurora islands in the southern Atlantic were reported, described, and located between the Falklands and South Georgia by Spanish sailors in the 1760s and 1770s, and then confirmed by another Spanish ship; but nobody else ever found them, including the scientific expeditions and the whale and seal hunters who circled the globe in the early years of the nineteenth century. These were the islands that Edgar Allan Poe's character was searching for in the *Narrative of Arthur Gordon Pym* (1838); but as far as we know, they simply did not exist. But then again, maybe they did, until they sank beneath the waves.

Some islands have an uncertain location but a precise place in cultural myth, like the island of Havaiki, the original homeland of the Polynesian peoples in a widely told creation story. It may be a mythical name for the real island of Raiatea in French Polynesia, or perhaps for a collection of islands widely separated in real space but closely connected in the imagination as places of original emigration for particular groups of people. In any case, in the words of one experienced observer, Havaiki requires respect as a symbol "of shared ancestry, of seafaring achievements, and of cultural antiquity." Describing islands in this way is a wise and realistic acknowledgment of their importance in the heritage and history of island peoples, and also serves as a reminder of earlier geological times when the islands had a different look.

Religious traditions, local or imported, have provided many kinds of island inspiration. Ophir, where King Solomon's legendary treasures came from—"And they came to Ophir, and fetched from thence gold [. . .] and brought it to king Solomon"—was known long before it was "discovered," though "undiscovered" people had been living there quite happily for a very long time. Originally presumed to be in Africa, and later in Asia, the European explorers in the sixteenth century believed Ophir to be in the Pacific—based both on a story that Solomon's ships took three years to reach it, and on a number of Amerindian accounts from the eastern coast of the Pacific telling of distant "treasure islands." European imperial and colonial mapmakers and politicians—and probably a few businessmen—duly located "some islands called Solomon" on maps before anyone actually saw them; and Gerardus Mercator, one of the best cartographers of his day, placed them about 10 degrees west of Peru, describing them confidently as gold-bearing—even though no European had ever been there. Then a Spanish explorer sailing from Peru reached what we now call the Solomons in the 1560s. He named one of the islands Guadalcanal, after his hometown in Andalusia, and the whole archipelago was named the Solomon Islands by an official reporting to the Spanish king. Then the islands were effectively lost to Europeans for nearly two hundred years, their location shifting with the winds and the waves. Now they have been found again; and although their connection with Solomon is more than uncertain, some residents of the islands still insist that they are his descendants.

The defining image of islands—surrounded by water, impossible to reach if you cannot float or fly, and protected from enemies, or simply from foreigners—has caught the imagination of countless storytellers, not least of them Shakespeare in his play *Richard II,* where he celebrates his "sceptred isle" as a "fortress, built by nature for herself / Against infection and the hand of war, / This happy breed of men, this little world, / This precious stone set in the silver sea, / Which serves it in the office of a wall, / Or as a moat defensive to a house, / Against the envy of less happier lands; / This blessed plot, this earth, this realm, this England." Such a blessed place offers both sanctuary from the wider world and a home for the fertile imagination.

The idea of a special space for ceremonies of belief, an island set apart from the everyday, is familiar from ancient times. Theater provides a secular example of such a space, which may explain why theatrical drama travels so well across cultures—everyone understands the idea of an island, and whether it be a space marked by lines in the sand or a carpet laid on the ground or a stage framed by seats or bleachers, the island form and function of a theater seems to be built into our consciousness. When we cross a theater threshold, just as when we cross the water to an island, we come into another world, a world which has its own customs and in which reality is transformed by the imagination. And truth in such a world, as Montaigne pointed out, may be a lie in the world beyond.

Biologists have expanded on Shakespeare's idea of an island as a "fortress, built by nature for herself" by defining an island as any habitat separated from another by a border; so a cluster of trees where a species of butterfly gather, for instance, would qualify as such an island, and so would a freshwater pond in the middle of a field. Sea or lake island habitats inevitably qualify, of course, but so do mountains where geology has produced rifts and ridges (mirroring those on the ocean floor) that make passage of certain plant or animal species difficult. This idea of "virtual" islands was developed two centuries ago by Alexander von Humboldt, the famous German explorer. (His brother, Wilhelm, was one of the first modern linguists to propose what is now called linguistic relativity, the notion that languages shape thought and feeling and behavior—in other words, that languages are like islands, joining individual "islanders" but separating them from those speaking other languages.) Alexander von Humboldt traveled widely, and he was one of the first to suggest that Africa and South America were once joined together. He also noted that cold, dry mountains had a similar climate to that of the low-lying arctic tundra, and that species that lived in each were similar, like species on islands with similar habitat. Darwin read his writings, and Humboldt's insights into the biological and zoological integrity of isolated habitats undoubtedly informed Darwin's theory of evolution. (The Humboldt brothers would not have been surprised to learn that the great diversity of species identified by Darwin and others on the Pacific islands is matched by an extraordinary linguistic diversity there. In the world overall there is, roughly speaking, one language for every million people. In the Pacific there is, again very roughly, one language for every thirty-five hundred people.)

There are other types of "islands," some of them so familiar that we forget their insular heritage. The so-called islets of Langerhans are groups of specialized hormone-producing cells in the pancreas, named after the nineteenth-century German pathologist who discovered them; and one of these hormones is insulin, the metabolic "island" from which the drug takes its name. The insulation we put in our houses and (as clothing) around our bodies, or that we rely on in an oven mitt when we pick up a pot off the stove, is there to create an "island" detached from its hot or cold surroundings.

Scientists often use island imagery in their stories. "Let x be such and y be so and so," they say—and we go with them to the island of algebra. Nuclear physicists refer to superheavy elements (which, at least theoretically, don't decay as quickly as lighter ones) as "islands of stability." Nobody has ever seen such an island, of course, which may be why it is referred to as the holy grail of physics; but like imagined islands throughout history, it has a special place in modern physical science. A massive particle accelerator called the Large Hadron Collider, with a seventeen-mile ring of superconducting magnets buried several hundred feet below ground (and bordering France and Switzerland), was completed a few years ago at a great cost to look for a subatomic particle—called a Higgs boson (or Higgs particle) after Peter Higgs, the British physicist who first proposed it—whose existence is crucial for the current standard story (that is, the model or theory) of the physical universe. Whence its nickname—the God particle.

Like earlier explorations, this scientific one holds promise of either confirming or contradicting our map of the world—in this case our map of the universe itself; and if it is successful, the scientists, like explorers for thousands of years, will undoubtedly use their discovery as encouragement to look for even more mysterious islands. The more real islands were discovered in ancient and medieval times, the more imaginary ones appeared on the maps of the world.

Humans seem to have sought out and given special status to islands, actual or virtual, since the beginning of time. Caves may have been the earliest of these, places which were isolated from the paths of daily activity and required special dispensations and directions to reach, often involving some danger. Islands in lakes and rivers followed soon after, and then islands offshore in the seas and oceans of the world. Oases and waterholes have invited island imagery since humans have made the desert their home, which is to say for tens of thousands of years; and the sacredness of circles—made of stone or wood—in many cultures has island implications, holding in as well as looking out and having no beginning or end, just like the allegorical promised land that St. Brendan recorded on his voyage. Mounds and mountains have often been described as islands. Some of them were man-made, such as the earthwork mounds of the pre-Columbian Cahokia people in the United States, and Silbury Hill, near Avebury in England. Monasteries were built from around the twelfth century CE on giant rock pillars at Meteora in Greece, islands in the sky that were only accessible by ropes and a net. Other island-mountain havens include Uluru (Ayers Rock) in Australia, which may have originated as a sacred site over fifty thousand years ago with those original seafarers, the Aborigines. The Black Hills of Dakota, sometimes called "an island of trees in a sea of grass," have had spiritual and secular significance to the Lakota Sioux for as long as memory holds. And such places find their way into our popular culture, from magic mountains to amusement park peaks, their appeal alternately dangerous and delightful. "Fostered alike by beauty and by fear," said Wordsworth of his upbringing in the mountainous Lake District of England, where lake islands are plentiful, many with the suffix "holme," after the Old Norse word for island.

It is this conjunction of the fearful and the beautiful, the chaotic and the well-ordered, the barbaric and the civilized, that has prompted people to build houses with fences and communities with walls—though houses themselves, all alone or grouped together, already serve much the same function. Bird nests and beaver lodges may be their natural counterpart, and the image of a castle surrounded by a moat hovers behind the notion that our home is our castle. And from ancient to modern times, from the Great Wall of China to the Berlin Wall, the building of walls around cities and countries has been a common response where there is no fortress built by nature, no wilderness of water to protect the community.

The early medieval monks who traveled to the islands in the North Atlantic in boats called currachs (framed of oak, with oxhide for the hull) embraced the extremes of wonder and dread, looking for both the sanctuary and the exile that they felt were necessary for salvation. This Christian fleet sought a desert in the pathless sea, pushing even the limits of North Atlantic island geographies. For instance, there was a monastic settlement founded in the seventh century CE on the lower part of a small island called Skellig Michael (Michael's Rock), on clear days within sight of County Kerry in Ireland. It was at the edge of human possibility; but it was surpassed by a hermitage built with uncanny skill and almost suicidal daring on a narrow pinnacle rising seven hundred feet above the sea on that same island. Skellig Michael was eventually abandoned, perhaps because climate change beginning around 1200 CE brought colder weather and frequent storms, making year-round living there impossible; but for a time this rock symbolized the ultimate island experience, exquisitely alone with both the natural elements and the supernatural spirits of place.

For thousands of years worldwide, spiritual callings have prompted people to set out to sea with very little food but a great deal of faith. In a paradox that fits with the experience of those marooned on islands, they had to maintain a sense of themselves as all too humanly ordinary in the presence of forces—material and spiritual—that were genuinely extraordinary. Otherwise, as Thomas Merton once said about the Christian hermits who retreated to the Egyptian desert beginning in the third century CE, they went mad. Many accounts of island travels are lost, along with the travelers; but a few wrote about their island adventures as though determined to maintain communication—even as they sought out island isolation. Some of the island sanctuaries were far distant, however, and those setting out for them said "good-bye" rather than "till we meet again," just like those who sailed in exile to Australia. But others were close to shore, requiring an imagination not unlike that of the small child who retreats to the backyard as a foreign country and a far refuge from family. Indeed, a surprising number of island sanctuaries—such as Mont Saint-Michel in Normandy; St. Michael's Mount in Cornwall; and Holy Island (or Lindisfarne) on the northeast coast of England, near Scotland—are tidal islands, joined to the mainland at low tide. Their spirit is exemplified by the island of Iona in the Inner Hebrides, just a mile from the much larger island of Mull; it was on Iona that Columba, a Gaelic missionary monk, first established a monastic settlement in the sixth century CE. Other sacred islands, like Inis Cealtra and Station Island, both in Ireland, are on lakes, while Mount Athos, in Macedonia, is on a peninsula but accessible only by boat. The water in all these instances seems to be less a barrier than a threshold, offering a rite of passage in which the traveler leaves the everyday world by stepping into a boat; and such a step has often been associated with voyages to the afterworld in religious traditions around the world.

Pilgrimages to holy islands have been widely practiced for a very long time; they include the annual Buddhist pilgrimage to the eighty-eight temples and shrines on the Japanese island of Shikoku and the Taoist and Muslim journeys to Kusu Island near Singapore, which, according to legend, came into being when a sea turtle saved two drowning men, one Malay and the other Chinese, becoming an island for them to rest on. But few islands have the mythical history and the hold on the imagination that Sri Lanka enjoys. *Sri* is an honorific, like master, and *lanka* comes from an ancient word for land; and Sri Lanka in ancient times could easily be reached by island hopping. Both the island itself and the small islands connecting it to the mainland of Asia have a unique place not just in the history of island trade and travel but in their association with several of the world's venerable religious traditions. Hindus say that the holy mountain Sri Pada on Sri Lanka is where Shiva stepped. Buddhists recognize the left footprint of Buddha on that mountain. Muslims believe it is where Adam first set foot when he and Eve fell from grace and left the garden with, as John Milton put it, "the World [. . .] all before them, where to choose / Their place of rest, and Providence their guide." Providence—or serendipity; and Serendip was the legendary name of Sri Lanka, a place of good fortune. (The English word "serendipity" was coined by Horace Walpole in the eighteenth century in his adaptation of a Persian fairy tale called *The Three Princes of Serendip*, who in Walpole's account "were always making discoveries, by accident and sagacity, of things they were not in quest of"—the quintessential island travelers.) So Adam and Eve chose the island of Sri Lanka—or it chose them—and it became their second paradise; Adam's name is recalled both on the peak of Sri Pada—Adam's Peak—and in that set of stepping-stones between India and Sri Lanka that is sometimes called Adam's Bridge. But since the Hindu epic *Ramayana* tells a story about how Rama built a causeway to the island, the eighteen-mile chain of limestone shoals and sandbanks is more often called Rama's Bridge.

Fortunate or blessed isles have decorated the maps and the minds of cultures all over the world. They are imagined realms, but we recognize them because they are inspired by the real world, and they can be discovered, or invented, almost anywhere. And just as different cultures have had different navigation and different craft for their seafaring, so these discoveries and inventions of favorite islands take a different shape in different times and places, embodying the dreams and desires that define humanity's best self (and sometimes its worst). The phrase "fortunate isles" was first given to the Canary Islands off the coast of North Africa by Juba of Morocco two thousand years ago; the Canaries were then lost to European chroniclers for well over a thousand years, until they were rediscovered in the thirteenth century. Like paradise, fortunate islands are forever being lost and found, in time and place "way back when" or "over the rainbow." But they *are* somewhere, sometime. Like the Antilles, whose name came from the Portuguese *ante* (opposite) and *ilha* (island) and referred both to a legendary eighth-century island of Christian refuge out in the Atlantic and more generally to islands across the wide water. Islands of good fortune. Islands of paradise or profit. Or island homes like William Butler Yeats's "Lake Isle of Innisfree," where the poet "shall have some peace." An old school textbook, which included Yeats's poem, suggested that "whether Innisfree is a real island or an imaginary refuge for the soul is of no consequence at all, so far as the spiritual significance of the poem is concerned." Maybe. But maybe it *is* of consequence, for reality is ultimately important. The imaginary Antilles may have been somewhere beyond the sea—but seafarers soon found, to paraphrase Gertrude Stein, that there *was* a there there. The islands of the Caribbean.

Ever since people have been telling stories—on all the continents; in the Pacific and the Atlantic, the Indian and the Arctic oceans; in all the smaller seas of the world; and on lakes and rivers inland—islands have provided a center of belief and a circle of wonder, a place where origins and endings are both real and imagined. This is the burden of one of the great poems of the English language, William Blake's "Jerusalem," where the question "And did those feet in ancient time / Walk upon England's mountains green?" is answered by a stern resolution not to cease, not to sleep, not to lose hope "till we have built Jerusalem / In England's green and pleasant land." The island that Blake invokes is both his and ours, both here and nowhere, both within sight and beyond the horizon. It is both a creation story, a story of the beginning—and a story of an end, an arrival. And it rhymes with a proposal by Noah Webster, put forward at the same time that Blake published his poem, to rid the American language of one of its British anomalies by changing the spelling of "island" to "iland." I-land. Home.

Afterword

From my home on the Pacific coast of Canada, I look out to islands. Their names hold the history, the humor, and the sometime hazard of islands everywhere. There is one a couple of miles offshore called Merry Island, where I paddle out to watch seals and sea lions lolling about the rocky shore, eagles circling the cedars and Douglas firs, and arbutus trees hanging out over the water with quirky grace and copper-colored bark. I also go because—well, because it's an island, over there across the water. On the south end of Merry Island, there's a lighthouse, inhabited by one of the few remaining lighthouse keepers on the coast. Its signal—one long and one short at an interval of fifteen seconds—shines into our house every night, gently lighting our living room and sternly warning the tugboats and the trawlers and the sailboats and the small motor yachts. All these vessels are negotiating their way north, past another island that is called Thormanby, and into the curiously named Welcome Passage—for the currents there swirl dangerously out of sync with winds that blow hard through the winter. This whole stretch of water is known as the Salish Sea, after the Aboriginal people who have lived here a very long time. They call Thormanby Island Sx̱wélap, and Merry Island Népshilin—though in 1860, officers on the British survey ship *Plumper* renamed the latter for James Merry, a Glasgow merchant who owned a horse they had bet on for the Epsom Derby that year. The horse's name was Thormanby; and the news that he had won was Welcome.

Whatever their provenance, island names are for sailors and storytellers to remember. The unknown and the unremembered are the greatest menace at sea; and names—of capes and coves and straits and narrows as well as islands—provide a kind of comfort, an illusion of the domestic in a wilderness of water. They are also reminders of earlier travelers, and the names change with seafaring sovereigns. The Salish Sea, for instance, is still listed on some charts as Georgia Strait, named for King George III. And names tell us that the British weren't the first Europeans around here. Behind Merry and Thormanby there's Texada Island, and further out the islands of Quadra and Valdes and Galiano, and they all recall the early Spanish sailors weaving their way among these archipelagoes. The Englishman George Vancouver—who visited Jamaica and Tahiti before surveying much of the Pacific Northwest—gave his name to Vancouver Island, which I can see in the distance. Along its coast, where cruise ships pass in summer on their way to Alaska, there is a narrow strait called by Captain Vancouver "one of the vilest stretches of water in the world." The reason was a twin-peaked underwater mountain, once an island but by the time Vancouver sailed through just ten feet below the surface and a menace to navigation. It was called Ripple Rock, and when I was growing up there were calls to take nature into our own hands and blast it out; but every attempt failed, until 1958, when one of the largest planned non-nuclear explosions on record blew it to smithereens. James Cook was on this part of the North Pacific coast, too, looking for the Pacific entrance to the fabled northwest passage; he is remembered in Cook Channel, across from the village of Yuquot where the whaler's shrine was located, while an island nearby is named after his companion on that trip—the notorious William Bligh.

Whatever their names, islands insist on themselves. Perhaps it is this seeming self-importance that raises the question with which I began this book. "What is an island?" The answer still stumps me, not least of all because although the map tells me that I live on the mainland, I can only get to my home by ferry boat or (recently) float plane. We are isolated from the rest of the coast by a deep fiord, surrounded by high mountains and steep ravines; and to further complicate things, we are also on a peninsula, with the ocean on all sides except for a narrow spit about a thousand yards wide on which sits a little town called Sechelt, an Aboriginal word that means "land between two waters." Sooner or later—and probably sooner, with the rising waters of climate change and the earthquakes that keep this part of the world uneasy—the sea will surround us and we'll be an island, and thus important, once again.

Islands give the world a human shape. "An island always pleases my imagination, even the smallest, as a small continent and integral portion of the globe. I have a fancy for building my hut on one," wrote Henry David Thoreau in *A Week on the Concord and Merrimack Rivers* (1849). So islands may represent both home and away, and our images of them often have to do with both arriving and leaving. Tourism nourishes the ambivalence, as it has since its inception. Creation stories and quest stories confuse it, which may be why they are the stock in trade of cultures all over the world. Sometimes we even come to the conclusion that there is no difference between coming and going. When we travel out to islands, we travel in to ourselves. When we look out to islands, we look in to the physical and natural and human history of the world. When we think about islands, we surrender ourselves to images of land and water in which neither has sovereignty and both have dominion.

As I look out to Merry Island and think about the island of Jamaica where my wife is from, I wonder whether islands may be less an invitation to the faraway than to the near-at-hand, less an appeal to leave than an offer to stay. Which would make the question "What is an island?" close cousin to "What is home?" Home is where you hang your hat, or where your heart is. It is the place you come from, or where you are going. It is where you choose to live, or where you have no choice but to live. Maybe the same place, or maybe not. Just ask Noah, or the woman who fell from the sky, or Odysseus, or Robinson Crusoe.

Notes and Acknowledgments

LIKE THE ISLANDS I write about, those who have helped me with this book are to be found both close to home and far away. Near at hand, my neighbors in Halfmoon Bay—Derek and Sue Hopkins and Harold and Liz Jones—deserve special thanks for watching as the winds blew, and knowing exactly when to wonder aloud. Fred Valentine and Gaither Zinkan, with many fine craftsmen, designed and built our home, from which we look out to islands in the Salish Sea. Britt Ellis and I started life puttering around Bowen Island, near Vancouver, and he has been with me—here and there—ever since. Tim Stewart asked me to sail from Stockholm to Copenhagen by way of the islands of Gotland and Bornholm; and thanks to him, we made it. George and Dianne Laforme have believed in me when I haven't, and I am deeply grateful. John and Barb Murray's friendship has been a blessing, pure and simple. And Ramsay Derry got me into all this writing business, way back when.

Many others offered insight into their particular islands: Tak Nakajima told me about Japan and the islands of the North Pacific; Liz and Bob Food showed me Salt Spring and the Gulf Islands; Dorik Mechau, Carolyn Servid, and John Straley brought me to Baranof in southeast Alaska many times, and Lowell Fiet and Maria Cristina Rodriguez to Puerto Rico; Curlie and James Lindsell first sheltered me in the British isles, and Stephen Regan returned me there. I traveled around Ireland with John Burns; and Paddy Stewart showed me what islands and gardens and theatres have in common. Eddie Baugh gave me an early welcome to Jamaica; Philip Sherlock, Rex Nettleford, and Jean Smith told me to get on with my writing; Elaine Melbourne offered her home and a place to write; Jo and Burchill

Whitman encouraged me in both clear and foggy weather; Denis Valentine gave me a sense of how to see the world the way a photographer does; Barry Watson showed me a painter's eye; Jeff Cobham credited me with something to say; Dian Watson called me cousin, and looked after everything; Barry Chevannes—the plainclothes Rasta—was a gift of grace; and the Goodison clan, bless them, took me in as a brother.

When I lived in England and learned the complicated habits of its sceptered and straitened islanders; when Peter Usher began writing about the Banks islanders in the Arctic; when Ian and Dave McDougall went to the Faroes; and when I traveled to the Caribbean and (a little later) to Fiji, the idea and the reality of islands beyond those I had grown up with started to take hold. My children—Sarah and Geoff and Meg—urged me on when I hit the doldrums, and cheered when I was underway again. Meg was a special help with geology, and got me to the Orkneys; Geoff with maps, and the surprises of sailing, which he handles with a seafarer's necessary nonchalance. David Smith kept me going with his healing arts and his stories; Brian and Linda Corman have been my friends for many years, waiting for me to settle down; Neil ten Kortenaar and Alan Bewell have supported me when they did not need to; the widely traveled Ian MacRae has provided insight and energy, as has Karen Mulhallen, who told me how to navigate around Venice; Michael and Rosalind Nightingale have always been there when I return, and so has Patrick Saul. Jerry Bentley has showed an interest in everything I have ever done, asking questions that remind me to look for answers; David Naylor, John O'Brian, Graeme Wynn, Richard Sanger, and Michael Dolzani assured me that someone was listening; Kevin Steels kept coming up with new islands; and for anything Spanish—indeed, for almost anything—I turned to Dan Chamberlain. There are many others whose writings and conversations I have learned from; I have tried to mention them all in the notes that follow.

Jan-Erik Guerth got me going on this maritime enterprise, and kept me company on what turned out to be a long voyage. He is a wonderful publisher and editor, with the focus of a ship's captain and the faith of a good friend.

This book is dedicated to the memory of my dear cousin Jack Cowdry—who drove big trucks and built small boats, knew everything about trains and could make anything work better than it did before he came along, including people—and to Rob Finley, whose knowledge of navigation and sailing craft is extraordinary, and whose writing about them beckons like the blessed isles. And it is written for my wife, Lorna Goodison, with whom I would travel to any island, and stay there if she would.

Thoreau's remarks in the Introduction and Afterword are from his book of travel meditations, *A Week on the Concord and Merrimack Rivers* (1849). The description of the Sundarbans is by the English writer Emily Eden, quoted by Michael Pearson in his article "Littoral Society: The Concept and the Problems" (*Journal of World History*, 17/4, 2006). Roman Jacobsen quotes the Majorcan storytellers (saying *aixo era y no era*) in his essay "Linguistics and Poetics" in *Style in Language* (1960), ed. Thomas A. Sebeok. The *Edinburgh Encylopaedia*, which is quoted at the beginning of each chapter, was edited—"conducted" is the lovely word used in 1830—by David Brewster.

In Chapter One, Philip Henry Gosse's *A Naturalist's Sojourn in Jamaica* is included in *Gosse's Jamaica 1844–45* (1984), ed. D. B. Stewart, along with Gosse's monograph on *The Birds of Jamaica*, and I have relied on both for my account of the flora and fauna of the island. Other sources include *A Photographic Guide to the Birds of Jamaica* (2009) by Ann Haynes-Sutton, Audrey Downer, Robert Sutton, and Yves-Jacques Rey-Millet; *Marine Life of the Caribbean* (2002) by Alick Jones and Nancy Sefton; *Trees of the Caribbean* (1980) by S. A. Seddon and G. W. Lennox; and Monica F. Warner's *Flowers of Jamaica* (2004). Allen Haaheim told me the makeup of the Chinese word for an island; and Hugh Hodges gets credit for reversing the metaphor for home, in his case with a Rastafarian turn inspired by the Jamaican elder Mortimo Planno. The details of Amerindian migration and settlement, and the Taino creation legend, are from *Taino: Pre-Columbian Art and Culture from the Caribbean* (1998), ed. Fatima Berch, with articles by Samuel M. Wilson, Ricardo E. Alegría, Marcio Veloz Maggiolo, and José Juan Arrom upon which I have drawn. Some general background is from *The Contemporary Caribbean* (2004) by Robert Potter, David Barker, Dennis Conway, and Thomas Klak, which also provided useful information on the geological and geographical character of the islands in the region. My crunchy-squishy description of the shore was inspired by Sean Kane's essay "Skaay on the Cosmos" (*Canadian Literature*, 188/ Spring 2006) about the Haida philosopher-poet Skaay, and the quotation from Plotinus in Chapter Two is also from this essay. The eloquent Salt Spring islander is the folksinger Valdy. The phrase "sweet burial" is reported in Richard Feinberg's *Polynesian Seafaring and Navigation: Ocean Travel in Anutan Culture and Society* (1988), which informed my description in Chapter Two of stars as songlines for Polynesian navigators, and what Feinberg calls their fundamentally optimistic attitude toward ocean travel. Wordsworth's tribute to Milton is from his poem "London 1802," and his line about the prison house of language is from his ode "Intimations of Immortality from Recollections of Early Childhood" (1807). Ron Hynes's refrain is from "No Change in Me," co-written with Murray McLauchlan; and in Chapter Five we hear another of Hynes's songs, "Sonny's Dream," which he graciously gave me permission to quote. Joseph Conrad's observation is from *The Mirror of the Sea* (1906), and Derek Walcott's from an interview with J. P. White in 1990. The quotations here (from "Silences") and in Chapters Two (from "The Ground Swell") and Five (from "The Fog") from E. J. Pratt are from *Complete Poems of E. J. Pratt* (1989), ed. Sandra Djwa and R. G. Moyles, copyright University of Toronto Press, with whose permission they are reprinted. Thurston Clarke's tribute to the savage sea is in *Searching for Crusoe: A Journey Among the Last Real Islands* (2001); and later (in Chapter Five), he brings Richard Dana (author of *Two Years Before the Mast*) into the story of Robinson Crusoe, and calls the musical *South Pacific* the *Robinson Crusoe* of the postwar years. John Donne's meditation is from his *Devotions Upon Emergent Occasions* (Meditation XVII). The version of the turtle island creation story retold here is from the Anishinabek—Algonquian-speaking peoples of the eastern woodlands of North America. The call on Katinanik and Katenenior is described in Thomas Suarez's indispensable *Early Mapping of the Pacific: The Epic Story of Seafarers, Adventurers and Cartographers Who Mapped the Earth's Greatest Ocean* (2004); and his discussion of early representations of the world (including the Islamic map "The Wonders of Creation") sponsored my own. Suarez also provides an entertaining illustration of the habit of putting home in the center of our maps; he tells the story of the elusive sixteenth-century Beach; and he recounts the Pohnpeian image of the horizon as the eaves of a house that I mention in Chapter Two. The stories from New Zealand and Hawaii are in F. Morand's *Legends of the Sea* (1980), trans. David Macrae; the Haida Gwaii creation story is told by Nang King.aay'uwans (James Young) in *Haida Gwaii: Human History and Environment from the Time of Loon to the Time of the Iron People* (2005), ed. Daryl W. Fedje and Rolf W. Mathewes; and Robert Bringhurst takes up the Haida tradition of creation stories in his magnificent translation of the epic recitation of the bard Skaay, *A Story as Sharp as a Knife: The Classical Haida Mythtellers and Their World* (2000). My discussion of myths draws on the work of Northrop Frye, to whom I am also grateful for the chicken and egg riddle, which he quoted (in a lecture titled "Some Reflections on Life and Habit" [1988]) from the nineteenth-century English writer Samuel Butler. The witty comment of Pliny about the antipodes is from Donald S. Johnson's *Phantom Islands of the Atlantic* (1994), which furnished me with all sorts of wonderful stories about imaginary islands, along with many of the images from the ancient world and the anecdote about medieval seafaring adventurers losing their civil rights on their return. Johnson also offers a description of the so-called Isle of Demons off the coast of Newfoundland, an account of the fanciful presence of Hy-Brasil and Buss island, and the mystery of the Aurora islands, which I recount later, in Chapter Five. His commentary on the story of Atlantis led me back to Plato's original invention in his late dialogues *Timaeus* and *Critias*; I also benefited from a review of Richard Ellis's *Imagining Atlantis* by Robert Eisner in *The New York Times Book Review* (July 12, 1998). The medieval Christian idea of islands as fragments of a fallen world is elaborated in John R. Gillis's essay "Taking History Offshore: Atlantic Islands in European Minds" in *Islands in History and Representation* (2003), eds. Rod Edmond and Vanessa Smith; and their introduction quotes the Tongan writer Epeli Hau'ofa describing the Pacific as a "sea of islands." Gillis's book *Islands of the Mind: How the Human Imagination Created the Atlantic World* (2004) was a source on many topics, including the old names for the surrounding sea, the geographical determinism of Ellen Churchill Semple, the invention of a seaworthy chronometer by John Harrison (in Chapter Two), and the curious imaginings of Iceland as Utopia (in Chapter Three). Along with others, such as Felipe Fernández-Armesto and Michael Pearson, Gillis reminds us that for millennia the Atlantic was populated by "shore" people, living on a watery frontier. This idea has been amplified for the Pacific by Greg Dening, whose writings about the natural features and human settlement of islands and the navigation of the Polynesian seas in *Islands and Beaches* (1980) and *Readings/Writings* (1998), and in articles such as "Geographical Knowledge of the Polynesians and the Nature of Inter-Island Contact" (in *Polynesian Navigation* [1962], ed. Jack Golson), have influenced almost everyone who has taken up the subject of islands in the past half century. The question of "colonist or castaway" is raised, in cultural and historical terms, by Mary Louise Pratt in *Imperial Eyes: Travel Writing and Transculturation* (1992); and Rebecca Weaver-Hightower provides some interesting commentary in *Empire Islands: Castaways, Cannibals and Fantasies of Conquest* (2007). Mark Patton has a good introductory chapter on features of island societies in *Islands in Time: Island Sociogeography and Mediterranean Prehistory* (1996). The Newfoundland fisherman landing on an island for the first time is quoted by Franklin Russell in *The Secret Islands* (1965), from which I quote Russell's own description of Funk Island in Chapter Four. The nineteenth-century anthropologist is Lewis Henry Morgan, and I was alerted to his comment by Graeme Wynn, who brought to my attention the comments by Rachel Carson earlier in Chapter One and Joseph Banks in Chapter Four. The lines from Homer are translated by Robert Fagles (*The Odyssey* [1996]); and Marvell's poem was called, simply, "Bermudas." John Masefield's poem is quoted with the permission of the Society of Authors as the Literary Representative of the Estate of John Masefield.

In Chapter Two, it will be apparent that the assessment of the seagoing craft and navigational skills of early Polynesian settlers has been hotly contested for much of the past century, with outsiders wondering whether such "primitive" peoples could possibly have reached the Pacific Ocean islands by anything other than serendipity. In the 1950s and 1960s, Andrew Sharp argued in favor of accidental rather than purposeful voyages, while others—such as the seafaring scholar David

Lewis—cataloged and demonstrated the techniques that would have been used by Polynesian navigators, including the observation of "sea marks." Lewis presented plausible descriptions of Polynesian outriggers and lateen sails in a chapter on "The Pacific Navigators' Debt to the Ancient Seafarers of Asia" in *The Changing Pacific* (1978), ed. Niel Gunson. Much of the debate was carried out in the pages of *The Journal of the Polynesian Society*, beginning in the 1920s with Elsdon Best and H. D. Skinner, with others (such as Greg Dening) joining in later; and their conversations have certainly shaped my own. This journal, which was founded in 1892, published Rua-nui's creation song of the skies, written down in 1907 by Teuira Henry; in 1894, Henry had presented the chant from Raiatea in the journal, describing it as coming "from the lips of Aramoua and Vara," written down by her grandfather J. M. Orsmond in 1817. The Samoan celestial catalog was published in the journal by John B. Stair in 1898; and in 1928, the litany of the nights of the moon was transcribed there by J. Frank Stimson from a performance by the great-granddaughter of Teuraiterai i Taputuarai, High Chief of Papara, Tahiti. David Olson illuminates the distinction between Polynesian and European navigation in *The World on Paper* (1994). The term "wayfinders" has had a long history, though its use by Wade Davis as the title of one of his books (2009) has given it new currency; and Davis's discussion of Polynesian seafaring (and of Malinowski's study of western Pacific sea travel) is as well informed as his many other chronicles of exploration. Patrick Nunn's writing about ocean islands has influenced contemporary island science and scholarship in ways that transcend the insular preoccupations of particular disciplines. I owe a great debt to his work, especially in books such as *Oceanic Islands* (1994) and *Vanished Islands and Hidden Continents of the Pacific* (2009). In the latter, he records the comings and goings of a number of vanishing islands and dredges up the story of sunken continents with the ease of an expert, offering a chronology of settlement on the Polynesian islands that incorporates the evidence of Lapita pottery and includes the grim consequences of climate change through the centuries, also describing the "islands" of palolo worms (mentioned in Chapter Five). The early twentieth-century ethnographer was Elsdon Best, quoted by Nunn (in *Vanished Islands*), where he also quotes the 1886 description by Shirley

Baker of Falcon Island (now properly called Fonuafo'ou) that is included in Chapter Three. Henry Menard's *Islands* (1986), a standard in the field, is graced by insight into the natural and human as well as physical history of Polynesia and many other ocean island regions, reminding us of the kind of connections Darwin made and other, more recent, recognitions, such as the success of random rather than targeted drilling in oil exploration. Jennifer Newell provides valuable information on trade and commerce in the Polynesian sea of islands, as well as about the livestock given to Tahitian chiefs, in *Trading Nature: Tahitians, Europeans and Ecological Exchange* (2010), framing her account with what she calls the "ocean-centric" attitude of scholars such as Epeli Hau'ofa; and along with Dulcie Powell in *The Voyage of the Plant Nursery, HMS* Providence, *1791–1793* (1973), Newell offers insight into the work of the indefatigable naturalist Joseph Banks. George Robertson's description of Tahiti is from his *Journal of the Second Voyage of HMS* Dolphin *Round the World (1766–68)*, while Bustamante is quoted in *Pioneers of the Pacific: Voyages of Exploration, 1787–1810* (2005) by Nigel Rigby, Pieter van der Merwe, and Glyn Williams. Bougainville's arrival on Tahiti and his descriptions are discussed by K. R. Howe in *Nature, Culture, and History: The "Knowing" of Oceania* (2000), where he quotes de Quirós on the islands of Vanuatu and the journal entry by Columbus (from *The Journal of Christopher Columbus* [1960], trans. Cecil Jane). Ian Cameron, in his *Lost Paradise: The Exploration of the Pacific* (1987), presents Bougainville's celebrated description of Tahiti, and in doing so opens up the question of the credibility of European seafarers' accounts. Cameron also has an informed report of the earlier seafaring of Dampier and Drake (and the windfall profits he secured for Great Britain, as calculated by John Maynard Keynes), of James Cook's travels and his use of Asian and Polynesian navigational experience in the Pacific, and of the depredations of scurvy during this period; he also provides the quotation from Jakob Le Maire. Much of the information about the plants and flowers of Tahiti is from Jean-Claude Celhay and Bernard Hermann's *Plants and Flowers of Tahiti* (1978) and Stanley L. Welsh's *Flora Societensis: A Summary Revision of the Flowering Plants of the Society Islands* (1998); and my discussion about a wider range of flora and fauna, including trade commodities, is indebted to Andrew Mitchell's *The Fragile South Pacific: An Ecological*

Odyssey (1990), from which I took the description of the overseas travels of the Tahitian birds and, in Chapter Four, the celebration of the unique species on the Hawaiian islands that would have appealed to Darwin. The marine life around Tahiti is described in detail by John E. Randall in *Reef and Shore Fishes of the South Pacific* (2005). Permission to quote the lines from "Island in the Sun" was kindly given by Irving Burgie. John Crowe Ransom's remark is from his book *The World's Body* (1938). The general commentary about navigators and navigations here and in Chapter Five is much indebted to Robert Finley, in particular his essay "Reading the View" (*Alberta Views*, July/August 2004) and his marvelous book about Columbus, *The Accidental Indies* (2000). For the story of *cyulis*, and indeed for many insights not only into the origin of words but also about North Atlantic island culture, I turned to Ian and Dave McDougall of the Dictionary of Old English project. The account of ancient tools on Crete comes from an article by John Noble Wilford in *The New York Times* (February 15, 2010). Pigafetta's map is included in Suarez (see above), along with his description of the "Unfortunate Isles" that I quote in Chapter Four, and of currachs in Chapter Five. Lionel Casson gives a lively account of Carthaginian and Portuguese travels in "Setting the Stage for Columbus" (*Archaeology* [May/June 1990]). Henry the Navigator's enterprise is elegantly recalled by Felipe Fernández-Armesto in *Before Columbus: Exploration and Colonization from the Mediterranean to the Atlantic, 1229–1492* (1987); and there are brief accounts of many of the early modern voyages of discovery in Peter Aughton's *Voyages That Changed the World* (2007). I am indebted to Donald Johnson (see above) for the seafaring origins of what we call disorientation, as well as the chronicle of confusions surrounding the measurements of the earth's circumference. The extent of the Dutch empire is outlined by Ernest S. Dodge in *Islands and Empires: Western Impact on the Pacific and East Asia* (1976), where he also discusses the trade in sandalwood and many other commodities and makes the suggestive connection between the old trade in salt and spices and the new commerce of oil and gas that I propose in Chapter Five.

In Chapter Three, Olaus Magnus's description is translated by Peter Foote (*Description of the Northern Peoples* [1996–1998]), and Saxo Grammaticus's words are from *The History of the Danes* (1980), ed.

Hilda Ellis Davidson and trans. Peter Fisher; both were provided to me by Andy Orchard. Halldor Laxness's genial revision of Icelandic history is from his essay "What Was Before the Saga: A Jubilee Discourse," published in the *American Scandinavian Review* (1974). Montaigne's quotation is from his "Apology of Raymond Sebond," translated by John Florio in 1603 and quoted by Michael Neill in his essay "The World Beyond: Shakespeare and the Tropes of Translation" (*Elizabethan Theatre: Essays in Honor of S. Schoenbaum* [1996]); and Auden's line is from his poem "Journey to Iceland" (*Selected Poems* [Vintage, 2007], ed. Edward Mendelson). Maclean's celebration is from his *Sketches on the Island of St. Kilda* (1838). My description of the earth's geology, along with hot spots, flow rates, and the closing and reopening of the oceans, owes a great deal to Menard (see above), who also came up with the image of the dip-stick; and I had considerable assistance with this chapter from Meg Chamberlin, who also told me about the Mozambique legend reported in *California Geology* (1996). "Times of darkness" are illuminated by Paul Taylor's article "Myths, Legends, and Volcanic Activity: An Example from Northern Tonga" (*Journal of Polynesian Studies* [1995]). The image of the wife of Makali'i is from Martha Beckwith's *Hawaiian Mythology* (1940). Stephen A. Royle has an entertaining account of Ferdinandea in his valuable book *A Geography of Islands: Small Island Insularity* (2001), with a good chapter on "Islands: Dreams and Realities," and an account of the flora and fauna of Ascension Island. The Scottish census definition of an island given towards the end of Chapter Four is from Royle, as is the school textbook reflection on Yeats's poem that I quote in Chapter Five. My mention of Tristan da Cunha was inspired by Joerg Jaschinski, who spent several years there with his family as medical officer. Darwin's observations in Chile are from his *Journal* during his voyage on the *Beagle*, as are his comments on barrier reefs; while his "truly poetical" conclusion is from a notebook entry written in 1838 when he landed back in England, and is quoted by Sandra Herbert in *Charles Darwin, Geologist* (2005). Robert FitzRoy's response to the earthquake at Conceptión is reported by Cameron (see above), as is the *Time-Life Nature Library* verdict on continental drift. The *Standard Encyclopedia of the World's Oceans and Islands* (1962) was edited by Anthony Huxley. The reaction of the people of Lisbon is described in memorable terms by the historian R. J. White in *The Age of George III* (1968). Cotton Mather's remarks are from his *Discourse Concerning Earthquakes* (1706). The sea areas listed are discussed at genial length by Charlie Connelly in *Attention All Shipping: A Journey Round the Shipping Forecast* (2004). Haldane shared the idea of "prebiotic soup" with the Russian scientist Aleksandr Oparin; and Stephen Jay Gould provides an informed and eminently readable history of geological science in *Time's Arrow, Time's Cycle: Myth and Metaphor in the Discovery of Geological Time* (1987). Keavy Martin reminded me of *nunataqs*, and the Inuit island inhabited only by women (in Chapter Five); and Benjamin Britten used the naval hymn in his opera *Noye's Fludde* (1957). Tuzo Wilson's career and contributions are recounted by his colleagues at the University of Toronto, and reported in the *Canadian Encyclopedia* (1985; online since 1999). The image of liquid honey is proposed by the Woods Hole Oceanographic Institution in Massachusetts. Bill Holm tells his island creation story in *Eccentric Islands: Travels Real and Imaginary* (2000); and in Chapter Five, Holm is the "modern writer" praising Prospero in Shakespeare's *The Tempest* and quoting John Fowles on the same subject.

In Chapter Four, Melville is quoted in Gillis's book (see above). Gillis also parses the word *travaille*, which was noted in the seventeenth century by Blaise Pascal. Darwin's remarks on the Galápagos are from his *Journal*, except for the diary entry, which is quoted by Nunn (see above) in *Oceanic Islands*; in *Vanished Islands*, Nunn describes the Wrangell Terrane, the Hawaiian lobelia, and the work of Alfred Wallace. The distinctiveness of the tortoise shell is mentioned by Cameron (see above), from whom I also take the image of a cultivated Garden of Eden. Menard (see above) offers a nice analysis of the role of chance in the settlement of oceanic islands, of the scientific conversations between Darwin and Wallace, and of the geological jigsaw that makes sedimentary rocks and granite the touchstone of continental formations; Menard is also the mariner who describes customary sea routes as turnpikes. My discussion of Darwin is indebted to many conversations with the late Richard Landon, Director of the Thomas Fisher Rare Book Library in Toronto and a serious Darwin scholar. The distribution of bats and insects is described in *The Pacific Islands: Environment and Society* (1999), ed. Moshe Rapaport, as are those curious Hawaiian raspberries and honeycreepers and the ocean-traveling, salt-tolerant coconuts. In *Islands of Fate* (2006), Fred Bruemmer identifies the flightless wood hen on Lord Howe Island and the miniature wooly mammoths on Wrangel Island, and he is the "astute observer" remarking on tame wild animals on the Galápagos. Bruemmer also describes the overbearing—he says "megalomaniac"—banyan, writes about the Diomede Islands in the Bering Strait, and portrays the Falklands as "the islands of birds, sheep and wars"; his account of the various species that fly or swim around particular islands is first-hand and first-rate. Michael Heppell gave me a remarkable description of the mangrove swamps on his approach to the island of Borneo (*Iban Art* [2005]). Information on some of the threatened or extinct species is from the *Endangered Species Handbook* (2005). The Faroese proverb "bound is the boatless man" is taken from the *Faroe Isles Review*, whose first few years of publication (beginning in 1976) provide a wealth of information about the islands, island travel, and the natural and human resources of island life in the North Atlantic. The Cornell Lab of Ornithology website was an invaluable resource for information about island as well as mainland birds. Christopher E. Filardi and Robert G. Moyle write about backtracking birds in the journal *Nature* (November 10, 2005). Radon Roosman has a fine article on "Coconut, Breadfruit and Taro in Pacific Oral Literature" in *The Journal of the Polynesian Society* (1970), on which I have relied for many of my examples. Wilde's comment about heredity is from his essay "The Critic as Artist" (1891). The description of Rockall is by James Fisher, who was on the British expedition that landed on the island in 1955. Derek Hayes, in his *Historical Atlas of the North Pacific Ocean: Maps of Discovery and Scientific Exploration, 1500–2000* (2001), gives a good account of whaling in the region; for the South Pacific, there is interesting information in William Dawbin's chapter on "Whaling" in *South Pacific Islands* (1984), ed. Peter Stansbury and Lydia Bushel; and, of course, there is *Moby Dick*.

At the beginning of Chapter Five, the chronicler (and a few pages later, the contemporary Newfoundlander) is Rex Murphy, whom I quote several times from his sardonic but sentimental celebration of Newfoundland on the fiftieth anniversary of its joining the Canadian confederation, titled "Between the Rock and a Hard Place" (*The Globe and Mail* [March 31, 1999]). Many of the terms used in the description of Newfoundland are from island friends, or the wonderful *Dictionary of Newfoundland English* (1982), ed. William J. Kirwin, John Widdowson, and George Story. Michael Crummey's poem is quoted with his kind permission. The version of "Wadham's Song" was taken from *However Blow the Winds: An Anthology of Poetry and Song from Newfoundland and Labrador and Ireland* (2004), eds. John Ennis, Randall Maggs, and Stephanie McKenzie, transcribed from *Old Time Songs and Poetry of Newfoundland* (1927). Oscar Wilde's comment is from his novel *The Picture of Dorian Gray* (1891); and Coleridge's from an article in the *London Morning Post* in 1801. Tennyson's lines are from his "Idylls of the King: The Passing of Arthur" (1869). D. W. Prowse's *A History of Newfoundland, from the English, Colonial and Foreign Records* (1895) establishes a baseline for the history of the island, and although it is about as straight as the island coastline, it includes some memorable documents and an account that is close to the heart of many Newfoundlanders, in much the same way that the sagas are to Icelanders. Mark Kurlansky's *Cod: A Biography of the Fish That Changed the World* (1997), on the catching, curing, and cooking of cod, is a lively source for gospel and gossip about traditions that shaped Newfoundland. The commercial success and the devastating collapse of the Newfoundland cod stocks are described in (among many other books) George Whiteley's *Northern Seas, Hardy Sailors* (1982); John Gimlette's *Theatre of Fish: Travels Through Newfoundland and Labrador* (2005); Pol Chantraine's *The Last Cod-Fish: Life and Death of the Newfoundland Way of Life* (1993), trans. Käthe Roth; *The Resilient Outport: Ecology, Economy and Society in Rural Newfoundland* (2002), ed. Rosemary E. Ommer; as well as in countless, though often sadly unreliable, government reports. Thomas Huxley is quoted by Ommer. The underwater survey of the fish around Newfoundland is from W. B. Scott and M. G. Scott, *Atlantic Fishes of Canada* (1988). Rousseau is quoted on *Robinson Crusoe* by Diana Loxley in *Problematic Shores: The Literature of Islands* (1990). Walcott writes of his picture of Crusoe in a manuscript draft of his book-length poem *Another Life*. Agatha Christie's musings on the purchase of an island are from *Come, Tell Me How You Live* (1946), a memoir about an expedition in Syria with her second husband, Max Mallowan, whose name she took for that book. I was alerted to her comment, and to much else, by Andrew Fleming in his essay "Island Stories" in *Scottish Odysseys: The Archeology of Islands* (2008), eds. Gordon Noble, Tessa Poller, John Raven, and Lucy Verrill, where Fleming proposes the phrase "desert in the pathless sea" that I use later. The account of Leicester Hemingway's island enterprise is from the Harry Ransom Center at the University of Texas at Austin, where his New Atlantis documents and memorabilia are housed. The "probably–probably not" call on the storytelling imagination is from Dan Yashinsky's incomparable *Suddenly They Heard Footsteps: Storytelling for the Twenty-First Century* (2004). The most complete account of the whaling shrine is by Aldona Jonaitis, with Richard Inglis, in *The Yuquot Whalers' Shrine* (1999). Andrew Stewart drew my attention to the description of the Isle Maree cure by Osgood Hanbury Mackenzie in *A Hundred Years in the Highlands* (1921); and he also told me about the Indonesian "hobbits" in Chapter Four. The catalog of islands exclusive to men and women, and the story of the Solomon Islands, owes much to Suarez (see above), as does their mapping. The biblical reference to Ophir is from 1 Kings 9:28. The "experienced observer" regarding the mythical island of Havaiki is Patrick Nunn (see above). Wordsworth's line is from Book I of his autobiographical poem "The Prelude." The story of Skellig Michael is told by Walter Horn, Jenny White Marshall, and Grellan D. Rourke in *The Forgotten Hermitage of Skellig Michael* (1990). Thomas Merton's comment is from his introduction to *The Wisdom of the Desert: Sayings from the Desert Fathers of the Fourth Century* (1960); Milton's description of Adam and Eve from Book XII of *Paradise Lost*; and Blake's "Jerusalem" from his preface to his epic poem *Milton*. Noah Webster's spelling is celebrated in H. L. Mencken's *The American Language* (1921).

In the Afterword, George Vancouver's line is from his *Voyage of Discovery to the North Pacific Ocean and Round the World in the Years 1791–1795* (1798). Andrew Scott provided me with the Aboriginal names for the local islands I refer to; his *Encyclopedia of Raincoast Place Names* (2009) is a treasure.

Index